ENCYCLOPAEDIA OF EMINENT THINKERS

VOLUME-XV: **C. RAJAGOPALACHARI**

Volumes by the same author

The Political Thought of ***Dadabhai Naoroji***
The Political Thought of ***Tej Bahadur Sapru***
The Political Thought of ***C. Rajagopalachari***
The Political Thought of ***Subhas Chandra Bose***
The Political Thought of ***Bhagat Singh***
The Political Thought of ***Annie Besant***
The Political Thought of ***C.R. Das***
The Political Thought of ***Lala Lajpat Rai***
The Political Thought of ***Sir Pherozeshah Mehta***
The Political Thought of ***Mahadev Govind Ranade***

About the Author

Dr. (Miss) Ashu Pasricha is a faculty in Gandhian studies at Panjab University, Chandigarh. She has many books to her credit including *Gandhian Approach to Integrated Rural Development; Public Administration: Discipline and Dimensions; Peace Studies: Discipline and Dimensions; WTO, Self-reliance and Globalisation; and Gandhi's Concept of State and Power.* Also, she has published a number of research papers in various Indian and foreign journals of repute.

ENCYCLOPAEDIA OF EMINENT THINKERS

VOLUME FIFTEEN

POLITICAL THOUGHT OF C. RAJAGOPALACHARI

Dr. Ashu Pasricha
Department of Gandhian Studies
Panjab University, Chandigarh

CONCEPT PUBLISHIING COMPANY, NEW DELHI-110059

Cataloging in Publication Data—DK
Courtesy: D.K. Agencies (P) Ltd. <docinfo@dkagencies.com>

Encyclopaedia of eminent thinkers / K.S. Bharathi.
v. 11-15 ; cm. — (Gandhian studies and peace research series ; 11)
Vols. 11, 14-15 by Ashu Pasricha; v. 12-13 by Jai Narain Sharma.
Includes bibliographical references.
Includes indexes.
ISBN 13: 9788170226840 (set)
ISBN 10: 8170226848 (set)

1. Statesmen—India. 2. Political science—India. 3. India—Politics and government. I. Bharathi, K. S. (Kuduva Swamy), 1956- II. Pasricha, Ashu, 1967- III. Sharma, Jai Narain, 1951- IV. Series: Gandhian studies and peace research series ; 11.

DDC 954.035 22

ISBN-13: 978-81-7022-684-0 (Series)
ISBN-10: 81-7022-684-8 (Series)
ISBN-13: 978-81-8069-495-0 (Vol. XV)
ISBN-10: 81-8069-495-X (Vol. XV)

First Published 2008

Published and Printed by
Ashok Kumar Mittal
Concept Publishing Company
A/15-16, Commercial Block, Mohan Garden
New Delhi-110059 (India)
Phones: 25351460, 25351794
Fax: 091-11-25357103
E-mail: publishing@conceptpub.com

DEDICATED
TO
THE SACRED MEMORY
OF MY COUSIN
(Late) MAJOR DR. ANUPAMA MEHTA
(1973-2005)

PREFACE

Rajaji—as he was popularly known—was the last of the crusaders, who were at the vanguard of the freedom movement. In his crowded life, he had distinguished himself in many fields and climbed the dizzy heights of 95 years on the ladder of age with an amazingly agile body and alert mind.

A multi-faceted genius of sparkling intellect, deep erudition, transparent honesty, bubbling humour, intense humanism, bold convictions and an unceasing zest for living a full life, Rajaji lent added charm to his colourful personality through his simple ways and austere habits. He had a rare combination of age and ardour, catholicity and conservatism, volubility and taciturnity, serenity and pugnacity; it was indeed a rare opportunity to watch him—his detachment in the midst of power, his religiosity mingled with a latitudinarian outlook—sometimes soft as a flower and sometimes hard as granite all these often showed him up as an enigmatic personality even to his old acquaintances.

Towards the end of his life, he became a great world figure. He attained this greatness as a result of a great powers of intellect, of devotion to great causes and the sacrifices he made in the interest of those causes. He was the first and the only Indian Governor-General of pre-Independence India and filled other high places in the official hierarchy with success and dignity. This was remarkable exterior greatness. Equally remarkable was the inner greatness of the spirit, the beauty of the personality that developed through the years. Selfless, noble, gracious and vibrant, this personality was more beautiful than ever before in the closing years.

He achieved eminence as a statesman and administrator, writer and scholar. His erudition in Sanskrit and ancient Indian lore enabled him to render into chaste English such masterpieces as the

Mahabharata and the *Ramayana*, enjoyed and appreciated by the intellectual and the illiterate alike, all over India. His deep knowledge of Tamil and its rich literature resulted in his translation into English of such classics as the Kural of Tiruvalluvar in his own inimitable style. He was also the author of original parables in Tamil in the manner of the *Hitopadesa*. He interpreted the culture and philosophy of India revealed in the *Vedas* and the *Upanishads,* the *Bhagavad Gita,* and other teachings of the ancient Rights and sages that have inspired and illumined generations of men and women from the dawn of civilization.

A lot has been written about the life and times of Rajaji, but the material on his political ideas is very scanty. In the present study, a modest attempt has been made to discuss his political ideas in a systematic manner.

Personal acknowledgements are no easy matters where so many have been kind. My special thanks are due to Professor R.N. Pal, Professor N. Radhakrishnan, Dr. K.S. Bharathi, Dr. A.D. Mishra, Professor Jai Narain Sharma, Dr. Meena Dutta, Dr. Seema Pasricha, Dr. Rashmi Sharma, Mr. Ajay Pasricha (Advocate) and Master Hemant Malhotra. Beyond that I can make only an all embracing salute to the galaxy of the scholars whose works I have consulted.

I am also grateful to Shri Ashok Kumar Mittal, my publisher, for his consistent zeal in the printing and publication of this volume.

Chandigarh **ASHU PASRICHA**

CONTENTS

Encyclopaedia of Eminent Thinkers Series

Vol. I. The Political Thought of **Mahatma Gandhi**/ *K.S. Bharti*
Vol. II. The Political Thought of **Nehru**/ *K.S. Bharti*
Vol. III. The Political Thought of **Rabindra Nath Tagore**/ *K.S. Bharti*
Vol. IV. The Political Thought of **Radhakrishnan**/ *K.S. Bharti*
Vol. V. The Political Thought of **Vinoba**/ *K.S. Bharti*
Vol. VI. The Political Thought of **Aurobindo**/ *K.S. Bharti*
Vol. VII. The Political Thought of **Jayaprakash Narayan**/ *K.S. Bharti*
Vol. VIII. The Political Thought of **Vivekananda**/ *K.S. Bharti*
Vol. IX. The Political Thought of **Ambedkar**/ *K.S. Bharti*
Vol. X. The Political Thought of **M.N. Roy**/ *K.S. Bharti*
Vol. XI. The Political Thought of **Dadabhai Naoroji**/ *Ashu Pasricha*
Vol. XII. The Political Thought of **Veer Savarkar**/ *Jai Narain Sharma*
Vol. XIII. The Political Thought of **M.A. Jinnah**/ *Jai Narain Sharma*
Vol. XIV. The Political Thought of **Tej Bahadur Sapru**/ *Ashu Pasricha*
Vol. XV. The Political Thought of **C. Rajagopalachari**/ *Ashu Pasricha*
Vol. XVI. The Political Thought of **Subhas Chandra Bose**/ *Ashu Pasricha*
Vol. XVII. The Political Thought of **Mohammad Iqbal**/ *Jai Narain Sharma*
Vol. XVIII. The Political Thought of **K.B. Hedgewar**/ *Jai Narain Sharma*
Vol. XIX. The Political Thought of **Bhagat Singh**/ *Ashu Pasricha*
Vol. XX. The Political Thought of **M.S. Golwalkar**/ *Jai Narain Sharma*

1

HIS LIFE AND TIMES

Rajaji's real name was Chakravarti Rajagopalachari. People affectionately called him Rajaji. How the family got the name of Chakravarti is an interesting story. In Sanskrit, 'Chakravarti' means the King of Kings. Once, when an ancestor of Rajaji was bathing in the river near their village, he saw a dead body come floating down the river. Fearing that if left so to float, vultures would eat up the body, he pulled it out of the river and cremated it. But, later, it became known that the dead body was that of a Harijan, considered by Hindus as an outcaste. Hence the Brahmins of the village expelled him. One day he had to perform an annual ceremony in memory of a dead ancestor. Such days are very sacred to Brahmins. But no Brahmin was agree to go to his house and partake of the ritualistic meal. Rajaji's ancestor was in great grief and anguish. Just then a person came along and said he was a Brahmin; he said he would perform the worship in that house and partake of the meal. As he was about to leave, he blessed the householder saying "Nallan Chakravarti" meaning a good Emperor. Then he vanished. The people who were present there thought he was not ordinary mortal, but a super human being. From then, this family got the name of Chakravarti family. So goes the story.

Rajaji was born on December 8, 1878 in Torrapalli village near Hosur in Salem District of Tamil Nadu. Torrapali itself was a small place situated on the bank of the Dakshina Pinakini. The family belonged to the Vadagale sect of the Srivaishnava community of Brahmins. His mother, Sringaramma came from Kuppam in Chittoor, in Andhra Pradesh. His father Chakravarti Iyengar was not only the village Munsif, but also a great scholar in the *Vedas, Puranas* and other Sanskrit lore. Chakravarti Iyengar had two brothers. The three brothers lived as one household forming what Hindu calls a Joint Family. Rajaji was the youngest son of Chakarvarti Iyengar. There were no sister, but a younger brother of Chakarvarti Iyengar had a daughter Namagiri, who was only a few months younger than Rajaji. The two grew in great affection together, much as if they were own sister and brother. This was in fact quite common and fully in the tradition of a joint family.

Rajaji was put to school in his fifth year after preliminary training at home. He proved to be an industrious student, prudent and anxious to learn. His school lessons interested him. All books attracted him. If he came across the prospectus of an insurance company, he would want to know what it was all about. His was an inquiring mind from childhood days. Very early his father had understood that Rajaji would always be questioning whatever he saw and did not understand answered all the questions of his promising son with apt awareness and exhaustive explanations. At times Chakravarti Iyengar had other problems engaging the mind; he would put aside his son's queries for some other time to answer. The lad understood the father so well that he never got grumpy when an inquiry of his was postponed for answer. He patiently waited for another hour to get the explanation.

Chakravarti Iyengar brought up his children in no luxurious ways; though he could afford to do it. From the outset, he wanted them to cultivate a taste of simplicity, which he himself had. In spite of the rather cold weather of Hosur, the boys were asked to sleep on the grass mats, with of course thick blanket cover. No bedsteads were provided. They had to study their lessons at night in the light obtained from the old style brass lamp, whose flame was fed by oil extracted from laurel nuts. Food he gave them in plenty. Milk, ghee and fruits were in the main menu besides rice, wheat bread and sweets. Even a few nickels were given them as pocket money occasionally.[1]

The Ancestors of C.R.

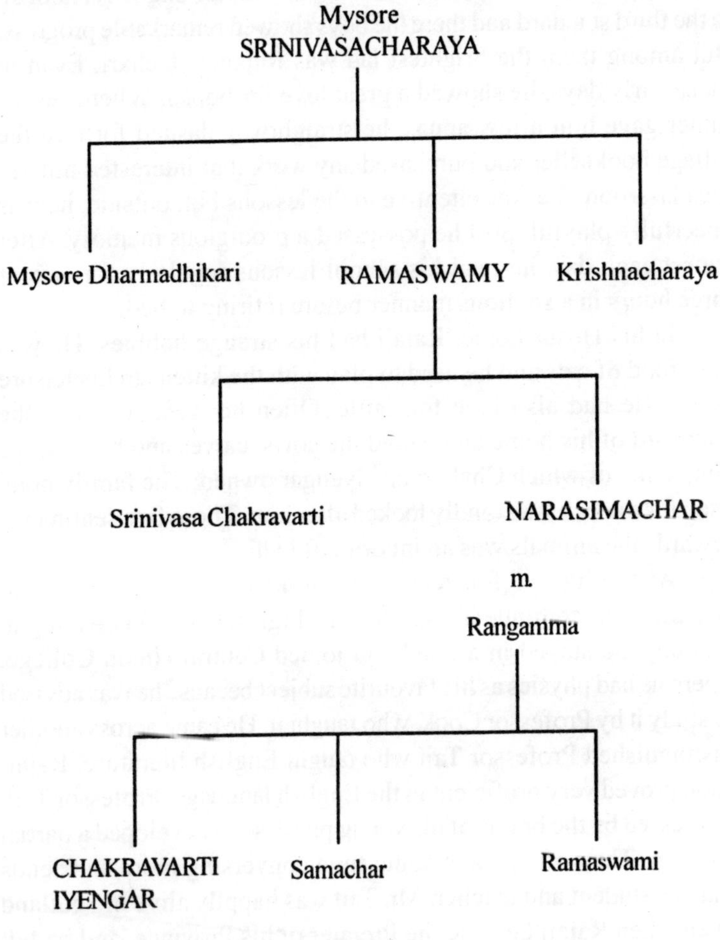

Chakravarti Iyengar's sons Narasimhachari, Srinivasa Iyengar, Rajagopalachari, all went to school Hosur had an English school up to the third standard and there the boys showed remarkable progress. But among them the brightest lad was Rajagopalachari. Even in those early days, he showed a great love for books. Whenever his father gave him a few annas, he straightway dashed forth to the village bookseller and purchased any work that interested him. In the classroom, he was attentive to the lessons but, outside, he was cheerfully playful. Still he possessed a prodigious memory. After sunset each day, he read his school lessons regularly for two or three hours in a studious manner before retiring to bed.

In his Hosur home, Rajaji had his strange hobbies. He was very fond of cats and he used to play with the kittens in his leisure hours. He had also love for cattle. Often he went round to the backyard of his home and patted the cows, calves and bulls in the pen, many of which Chakarvarti Iyengar owned. The family pony used for a coach was fondly looked after him. These fine sentiments towards the animals was an inborn gift in him.

At the age of fourteen, Chakarvarti Iyengar sent him to Bangalore to continue his studies in a high school. On arriving in that city, he stayed in a hotel and joined Central Hindu College. There he had physics as his favourite subject because he was advised to study it by Professor Cook, who taught it. He came across another distinguished Professor Tait who taught English literature. Rajaji soon proved very proficient in the English language. Professor Tait, impressed by the brains of his young pupil, soon developed a partial for him. They often spent some time conversing more as friends than as student and teacher. Mr. Tait was happlly alive in Scotland even when Rajaji became the Premier of his Province, and he felt supremely happy on knowing the greatness Rajaji had attained for himself in life. Tait died in 1945. But the pupil even then held him in grateful memory. Rajaji said of him:

"How can I forget Mr. Tait? It is not a mere figure of speech to say that he was a father to us all in our college days."[2]

Professor Tait once advised Rajaji to proceed to England and study at Oxford. But he was not enthusiastic about the suggestion. He politely declined saying that "India is quite enough for me".[3]

Rajaji's days in Bangalore were eventful. Besides cultivating the acquaintance of men of learning like Tait and Cook, he got popular with a large circle of friends, many of them well known in Mysore. Even when young, he was popular with his associates. He had then no outer arrogance and vain presumptions, which would have made him an object of detestation; instead he was distinguished by an amicably piquant temperament and this made him a universal favourite.

He was perceived as a model student in school. He scored enough marks in all subjects and shone well as a student of English literature. He mastered all Shakespearean works and ceaselessly read the novels of Sir Walter Scott. Modern English games were not popular in India in those days and the seeds of them were in fact just being sown here. At any rate, Rajaji had no interest in them. But he went out for long walks in the evening, walking always in a thoughtful and serious mood. Sometimes, he had a friend or two to accompany him. But his constant companion in those days was Mr. Navratana Rama Rao, a former Civilian Officer in Mysore. Having the ideal of simplicity before him, he proved to be a thrifty student. The help he got from his father each month was just a small amount of money but enough to cover his essential expenses. Besides paying the school fee and boarding charges, he invested little on clothing. He was content to have the bare minimum articles of apparel which were necessary for day to day living. He no doubt wore a *dhoti*, black coat and cap but that was all. Yet, his personality had a shine of polished marble. His whole idea was simplicity, again simplicity, in clothes and mode of living.

Rajaji, once speaking at his old college in Bangalore went into a reminiscent vein. "I think I have spent my happiest days in this college", he told a distinguished gathering present to hear him. Then after a few years of study in the Central College in Bangalore, he moved for a two-year Law Degree to Madras. That brought his college days to an end.[4]

At twenty, Rajaji had become a lawyer, ready to throw himself into the bread-winning struggle. Why he chose law for a career? He must have been attracted to it only because of the real spirit of the Bar. The arduous struggle, the blows given and received, the exultation of victory, the sting of defeat, which are lawyers' daily experiences, far from breeding division and ill-will, only bind them more closely together by the ties of comradeship for which one would look in vain in any other arena of the ambitions and rivalries of men. Moreover, he could see the famous men of those days in South India had been invariably members of the legal profession. Old Chakravarti Iyengar, himself a Government servant at Hosur, wished his son to pass the Indian Civil Service examination and become a District Collector. But Rajaji had no flair for the glories of a civilian office. He would rather be independent man with an independent income. He would not be a servant in an administration run by an alien race.

When he was studying in the Law College in Madras, an interesting incident happened.

Swami Vivekananda visited Madras. He was put in the very hostel where Rajagopalachari was staying. Going round the rooms of the students in the hostel, he entered Rajagopalachari's room. He saw on the wall a picture of Lord Krishna. He asked, "Why is Lord Sri Krishna blue in hue?" Rajagopalachari, still a student, answered: "Sir, the sea is limitless. So is the sky. And both are blue. God is also boundless. And so His hue is also blue."[5]

Swami Vivekananda was overjoyed. He said the young boy would rise to eminence and would become very famous.

To begin his career at the Bar, Rajaji looked at Salem, the chief town in his native district. So he went there. It was a wise decision, for he thought he would build up a practice there first before "invading" Madras for legal eminence. To aim at a lucrative practice all at once in Madras itself would be a folly, he had realized. He also understood that the Madras lawyers who earned thousands a month got all their money through appeal cases sent to the city from the district centres. Giving due thought to these facts, Rajaji went away to Salem.

When he arrived in this town to commence practice, his aged father Chakravarti Iyengar and the brothers Narasimhachari and Srinivasa Iyengar too came to live with him. His mother Singaramma was no more alive, she died in the meantime. Hiring a house in the Hasthampetai extension area in Salem, Rajaji commenced his works in courts. He looked up to some senior lawyer to initiate him in the profession. He started independently. Of course veterans of the Salem Bar like C. Vijayaraghavachariar were there and with them he picked up an acquaintance.

Pretty soon he became a 'star' lawyer in the criminal courts. He was so skilful that he won even very complicated cases quite easily. On the civil side, he had few cases. He prepared wills and helped clients to take out letters of administration and so on. The Shervaroyan Hill Planters on hearing about his abilities engaged him as their legal adviser. His fame soon spread. He conducted very difficult cases independently. His knowledge of law, his intelligence and his fearlessness, brought him great fame and wealth in a short time. He became very famous as an advocate in Salem. The landed aristocrats in Salem now sought his services and to him they entrusted all court work. Then his part in a case connected with the Mittadar of Thiruchengode enhanced reputation in the legal circles.

With the steady rise of his fame as an efficient lawyer, his fortunes too rose in proportions. With the increase of work, he moved on to the second Agraharam Street so that he might be nearer to the courts of Salem. His prosperity as a lawyer could be imagined when we come to know that even within two years of his going to Salem, he purchased a handsome horse-coach. Later, he was the first man to own a motorcar in Salem, himself investing about three thousand rupees to purchase a French Diarec car. He dressed himself well. In those days he was clad in a *dhoti*, black alpaca coat, lace turban, socks and shoes to match.

Neither the fame as a sound lawyer, nor the admirable monthly income he earned thereby, ever made Rajaji self-satisfied. He continued to be a hard worker. From sunrise to sunset he had before him legal work. Towards evening he used to go to Salem Literary Society, an institution of which he was one of the founders. There, he met his colleagues in the profession and exchanged views. But he was not extravagant with words. "He was somewhat a reserved sort of man in those days" said his brother Srinivasa Iyengar.[6]

Soon after he came to Salem, Chakravarti Iyengar insisted that Rajaji should be married. Messengers were sent out to look for a bride. But when willing fathers who had daughters to be married heard about the bridegroom as a bespectacled bachelor, their faces turned pale. They did not want their daughters to be given to any sight-defective man. However, Vasihnavite Brahmin villager of Chittoor never worried about these silly considerations when approached and he readily agreed to get his daughter Shri Alamelu Mangammal married to the lawyer from Salem.

Alamelu Mangammal knew Tamil and Telugu well and was an intelligent young lady. She was a model of devotion to her husband. In return, Rajaji showed great affection for his wife whom he tenderly looked after. Their life together was happy for a time. She gave birth to two daughters and three sons. As years went by, she developed physical ills and became weak day after day. In 1917 her illness unhappily took a serious turn. Rajaji now left the Aghraharam and moved on to the Masonic Building in the Suramangalam area to reside. Able medical aid was given to her. He spent sleepless nights nursing his ailing wife. He gave her effective medicines and helpful diet at hours prescribed by the physicians. He sat by her bed, offering words of consolation. In spite of all this care and devotion shown in nursing her, Alamelu Mangammal passed away. Till death came, Rajaji was care-worn and worried. But when it finally did come, he faced the event courageously, for he knew well that there was nothing to worry about beyond the end. After eighteen years of married life happily spent he lost the comrade of comrades, the sharer of joys and sorrows, hopes and fears, the fond and merry companion who was always a crown to her husband. But then his worry was about the five children now. Though one or two of them had attained boyhood, the rest were infants.

When he was just new to his practice, the Government had prosecuted a patriot by name Varadarajulu Naidu; the charge was that he had spoken against the Government. Rajagopalachari was the advocate for Varadarajulu Naidu. Whenever Rajaji stood up to speak, the judge would say, "Please sit down." One day, at the very outset Rajaji stood up and made a submission: "I have great respect for the Honourable Judge. But if like a teacher in a classroom, Your Honour always asks me to sit down, I shall not be able to discharge my duty; and it will not add to the dignity of this court."[7] The Judge apologized to Rajaji. Finally, on the basis of Rajaji's arguments, Varadarajulu Naidu was acquitted.

In Salem, Rajaji made a mark as an important citizen. He developed new ideas in him and whenever he found that social reforms were opposed to the age-long Brahminic traditions, he was inclined to forget traditions and face facts. Chakravarti Iyengar, his father, was of an old type. He did not like his son becoming one who would break traditions with new ideas about men and matters but then he was not the man to interfere seriously. Rajaji had the greatest reverence for his father throughout but then he had his own strong, free will too. Soon he found interest in social reform work and taking a liking to master the political affairs in the world. A few years after coming to Salem, he was elected Chairman of the Municipality. In that office, the Council Hall, hangs his picture even today as a mark of gratitude for his good work.

Rajaji believed that the individual was part of socicty and must make his contribution to social life. In 1917, he became President of the Municipal Council in Salem. No sooner did he become President than the work of the local body got a new impetus. He started adult education classes and also night schools for factory workers. Schools for Harijans were also opened. Rajaji appointed a committee to prepare a glossary (list of difficult words) in Tamil so those science subjects could be taught in that language. He also drew up a plan for the supply of water to Harijans. A Master Plan for the development of Salem was also formulated.

Rajaji did much to better the lot of the Harijans. When he admitted Harijan students into a hostel, even some elders whom he respected very much opposed him. But Rajaji did not relent or retreat.

Conscientious to duty, he was in the Municipal Office by eight in the morning, as though he was paid for the work! He also put to shape his ideals of equality of man in public while holding the office of Municipal Chairman. But in attempting to do that, he encountered with many stout oppositions. For instance, he had *pariah workmen* to open the Municipal water supply taps in the Brahmin inhabited streets. The high-caste folk vehemently protested. The whole story as fell from his lips: "Less than 20 years ago there were two Panchama employees in the Municipality of Salem. Their work was to open and close the stop-cocks for the water taps in certain quarters. The Engineer of the Municipality transferred these two men in the ordinary course to the Agraharam ward. I was the Chairman of the Municipality, but I did not happen to know this when it happened—not that if I had known it, I would have asked the engineer not to do it. Then there was a great agitation in the Agraharam and all the Brahmins were angry with me. 'You are a social reformer, but you have no right as Chairman to force your views on us. You have put these untouchable employees on to the Agraharam merely to worry us. There are so many places where you could have put them.' I had numerous friends all over the Agraharam, all co-workers in various activities; and one of them said, 'My old mother is objecting to use the water. Help me.' Another said, 'My aunt has not bathed.' And a third, 'Do you wish to kill my grandmother? She has not eaten at all for two days.' All the widows cast looks of hatred at me as I passed along. I knew how they loathed the touch and contamination of an 'Untouchable'.

"Now it was lucky that the Mahatma had not promulgated his *Satyagraha* in those days. If they had known it then, they would have used it against me straightway. They would have all gone on hunger strike. They had a genuinely felt grievance and would certainly have given up their food with ease and with a sense of complete martyrdom. What would have been my fate? Even the Government would have telegraphed to me to withdraw the 'Untouchable' from the Agraharam ward. Government would have told me. 'Your social reform is right but you do not know your people,' and would have ordered the transfer of the Untouchable stop-cock men over my head.

"Luckily, however, *Satyagraha* was not known to the good people of the Agraharam. So they did not go beyond verbal objections and moving propositions in the Council. They argued but I was obdurate and said, 'you may have your religion or whatever it may be in your private capacity. The Municipality cannot make distinctions of caste. I cannot displace them. Today you ask for transferring them. If I transfer them to another ward, the people of that ward will also ask for a transfer of these Untouchable men. It is better that you bear with them.' They had no answer. They dared not suggest dismissal. After some time everything subsided. I too forgot all about it until some time later. During the reforms enquiry, before the Parliamentary Joint Committee, it was said, 'There is no chance of Democracy progressing in India with its innumerable castes and distinctions.' Then the late Mr. K.T. Paul who was a Salem man and who was there said, 'One Municipal Chairman, a Brahmin, did succeed in fighting caste distinction in his Municipality. India is changed.' When I was told of this I was reminded of the affair and when the Brahmin friends of Salem who had objected to the employment of the Untouchable men heard of the Joint Committee reference, they too were glad and thought it was rather good I had been obstinate."[8]

Secondly, we see him, at this stage, giving scholarships to Untouchable students. But the school authorities refused to admit those boys. Rajaji insisted and finally got them in safely. Such a noble attempt he had made even in the years when crusades against Untouchability were unknown in India. They clearly mark out the man for his far-reaching ideals of social equality and large-heartedness.

For two years, Rajaji was the Municipal Chairman and his good work during that period stands out as a matter of merit even today. Commenting on the Municipality's administration report in those days, Mr. E.W. Leigh, the then Collector of Salem wrote that he wished all the Municipalities in the Province were maintained as that of Salem. For Mr. Leigh's judgment, I pay the highest regard because, I chanced to know him in Madras during his last years. After retiring from I.C.S., he did missionary work and was in-charge of the Church of Scotland Mission in Madras. We often met in his bungalow in Ritherden Road. When Rajaji became Premier of Madras, Mr. Leigh was there to compliment him.

This energetic lawyer who found himself in a network of activities in Salem soon became a victim to that dreadful disease Asthma. It troubled him often. He was his own physician to cure it, he once said. During the First World War years, he was once seriously taken ill. One early morning he set out on a trip to

Coimbatore to attend to legal work. He rode up to the railroad station in his Diarec coach-model motorcar. With him was Mr. Narasimha Iyengar, his junior counsel. They had the hood of the car folded up and the chill weather badly affected Rajaji. During the train journey he felt feverish and returned to Salem from a mid-way station. The doctors diagnosed pneumonia. For several weeks he was forced down to bed. Efficient medical aid was summoned. Even then, the doctor gave up hope. But Rajaji had the will to live. He changed his doctor immediately and was soon on the road to recovery. He got well soon and was restored to normal health once again.

Courage is yet another commendable quality in this veteran's lawyer, admired by the people of Salem. When at the top of his legal career, he one day came out of the Court and boarded his horse-coach. The driver was just then away. But as soon as Rajaji got in, the spirited horse took to his heels. The animal ran at a terrific speed. People who were nearby shouted out at Rajaji asking him to get down but he would not. "I will see where all he goes" he retorted and stuck to his seat inside. After running to a length of six furlongs the coach struck a stone by the side of a river and stopped. Rajaji calmly got out as if nothing had happened and gently patted the animal. The terror-stricken spectators of this incident admired him for the remarkable boldness he exhibited.[9]

A similar incident occurred years later. He had an abscess to be removed from his thumb. It was stated to be rather serious. His doctor advised him that only under chloroform could he operate. Rajaji did not agree. Finally, the operation went on without chloroform, and the patient courageously stood up to the painful ordeal for over an hour remarkably well started taking interest.

Besides social reform work, Rajaji in politics during his sojourn in Salem. At the Literary Society, he used to play tennis in the evenings and then sit on a bench on the lawn to discuss current politics with his friends. He read newspapers and digested the day's news with care. He had a keen eye on foreign news too and thus, he studied the happenings in the world abroad. He then heard about the Passive Resistance Movement, which Gandhiji was leading in South Africa. He was wonder-struck by the Movement. Now, Rajaji admiring Gandhiji even in those days, collected funds in Salem and sent them to him in South Africa. Later on, Madame Annie Besant roused the Indians to fight for their Home Rule rights more ardently, Rajaji sympathized with her and lent his support to the cause.

REFERENCES

1. Perumal, Nilkan, *Rajaji* (Calcutta: Maya Publications), 1948, p. 14.
2. *Ibid.*, p. 16.
3. *Ibid.*
4. *Ibid.*, p. 18.
5. Gandhi, Rajmohan, *The Rajaji Story: A Warrior from the South* (Bombay: Bharathan), 1978, p.11.
6. Perumal, Nilkan, *Rajaji, op. cit.*, p. 21.
7. Iyengar, Masti Venkatesa, *Rajaji* (Bangalore: Jeevan Karyalaya), 1975, p. 10.
8. Perumal, Nilkan, *Rajaji, op. cit.*, p. 27.
9. Gandhi, Rajmohan, *The Rajaji Story: A Warrior from the South, op. cit.*, p. 24.

2

IDEOLOGY AND PERCEPTION (RAJAJI AND MAHATMA GANDHI)

Though Rajaji had met Mahatma Gandhi for the first time in 1919, he had been getting news about him from the newspapers since 1906. To quote his biographer, "From 1906 onwards newspapers had published items about a Gujarati barrister called M.K. Gandhi who was battling there for the rights of Indians. The accounts impressed C.R. An Indian deported from South Africa visited Salem—If I remember right, it was in 1903—stayed a while in C.R's home, and told a group, Gandhi is small in size but his heart is bigger than the Shevaroys seen from Salem.

"Then C.R. picked up, by chance, from the desk of his friend Narasimha Iyer, Civil Disobedience by Henry Thoreau. He found the book "very interesting". It seemed to supply the theory for what Gandhi was practicing. Soon another book made an impact—Gandhi's Indian Home Rule, banned in South Africa sent to C.R. by a deportee he had met in Kurnool in 1910. Reading the book with very great avidity, C.R. was struck more by Gandhi's freedom from timidity and his refusal to recognize the English as superior than by his defence of non-violence.

"Indian Home Rule was written as a dialogue between the author, cast as 'Editor', who advocates a non-violent struggle, and a friend, described as 'Reader', who often argues as a believer in violence. It seems that the book was first reviewed in India by C.R. In his review, of which only a sentence has been traced, C.R. could not resist a dig. "Naturally," wrote C.R. "Editor is a better debater." The implication was that the case for violence had not been presented well enough.

Not yet convinced that the use of violence was wrong, C.R. even thought that Gandhi might be induced to join the revolutionaries and, with his qualities, lead them to success."[1]

Rajaji in those days was more sympathetic towards revolutionaries. He was hopeful that Gandhi might give a lead to them when he came to India. After the arrest of Tilak and later of Khudiram Bose, who had thrown bomb on a British Judge in Muzzafarpur, the revolutionaries, it seemed, had lost the day.

The Moderates were happy because of Minto-Morley reforms, which contemplated setting up of an advisory council of notable Indians nominated by British officials. But soon the Moderates were also disappointed because Morley had declared that the new scheme is not intended to lead to a parliamentary system in India.

It may be mentioned that Rajaji had already warned his countrymen against Minto-Morley reforms. According to him such an assembly "will be a fresh weapon in the hands of those whom we fight…now, though very reluctantly, calling our enemies." In the *Hindustan Review* of July 1907 he called these reforms "deceptive and cleverly divisive." He urged his countrymen not yield to "small temptations in the moment that decides."[2]

Though Rajaji had his leanings towards revolutionaries, yet he liked the Gandhian concept and in fact applied it in Salem before he met Gandhi. The Superintendent of Police, and Englishmen had invited leading Salem citizens to a sports meet. But there were separate enclosures for the British and the Indians. He was annoyed to learn it. He not only asked his friends to walkout but also himself led a walkout of all the lawyers.

In 1912, bomb was thrown on the Viceroy, Lord Harding at Chandni Chowk. It again revived people's interest, though for a very short time, in the revolutionaries. Rajaji was still hopeful that Gandhi might come to India and give a lead to the revolutionaries. His belief was strengthened because of a discussion in Salem about the future of revolutionaries in which Dr. T.S.S. Rajan had asked the people to wait till Gandhi came to India and gave them a lead. Rajaji felt that Dr. Rajan had a point.

But he realized in 1913 that Gandhi would be the last man to support revolutionaries. An article by Gandhi had appeared in

Modern Review of that year giving account of Gandhi's three terms in jail. Just at the time Rajaji read J.J. Doke's book on Mahatma Gandhi; he was convinced that Gandhi would never support violence.[3] In the words of his grandson, "If C.R's mind was çleared of a false hope, his heart was stirred by the picture of the suffering an evidently just man was inviting on himself. It was moved afresh when he heard, in September 1913, that Gandhi's wife and young sons had been arrested for peacefully opposing a racial tax.

"He reprinted Jail Experiences with his own money and, indeed, his own labour. Standing in front of the treadle, treading the machine himself, he took off the sheets for the booklet. C.R.'s introduction, in which he asked for funds, showed how he had been affected.

"'Shall we', he wrote, 'sit happy on our homes, or shall we give only our tears? It is not given to all to exhibit the strength of M.K. Gandhi. He must be ranked with the avatars....'

"Let us give up perhaps a few luxuries and support them in the holy battle they have resolved to fight once again. It will chasten our characters (and) bind us by a fresh and direct tie with those brave countrymen of ours...."

"Collecting fifteen hundred rupees, C.R. sent the money, with heartfelt reverence, to Gopalkrishna Gokhale, the Poona leader for whom Gandhi had a deep regard. Gokhale forwarded the sum to Gandhi.

Though Gandhi's struggle seemed confined at the time to South Africa, its reports suggested to him a weapon which we can wield...in India."

Gandhi had succeeded in South Africa but will the new weapon of *Satyagraha* work in India also? Rajaji was among the first who was sure that it would. He clearly spelt his views in a paper entitled *M.K. Gandhi: His Message to India.* "Led by him, our brothers and sisters of South Africa have so acted that Indians may forget their unworthiness for a time and walk proudly in the world."[4]

Somewhere he had acquired a portrait of Gandhi and Kasturba; it was while presenting it to the Salem Literary Society that he had read his paper.

Rajaji had so much regard for Gandhi that he told his ten-year-old daughter Papa not to use *Avan* while speaking of Gandhi.

Avan in Tamil is "he" without respect. "Why not?" asked Papa "Is he a Brahmin?" "He is not a Brahmin", Rajaji replied "but he and his wife are like Rama and Sita."

It was in these circumstances that Gandhi finally returned to India in January 1915 at the age of forty-five with little money but with a burning desire to serve his people. Though many people had heard of his achievement in South Africa with his new weapon of *Satyagraha,* he did not attract crowds in Bombay on his return to India. While Gandhi was still in South Africa struggling for his countrymen with his new weapon, *Satyagraha*, people in India had begun to call him Mahatma an appellation, which struck to him firmly and became a part of his name. On reaching India, Gandhi met his "political guru", Gokhale, and told him that he would spend the first year in India studying the country, with "his ears open but his mouth shut."[5]

Gandhi treated all those who lived at Tolstoy Farm and in Phoenix Settlement in South Africa as members of his family. Since many of them had come to India, Gandhi needed a home to house the large "family". The Society of the Friends of India supplied the money necessary for the purchase of land, and the building of huts.

Gandhi founded an ashram, in May 1915, on the bank of the river Sabarmati, on the outskirts of Ahmedabad. He called it Satyagraha Ashram but it came to be known, from the place where it was built, as Sabarmati Ashram. The inmates, about twenty-five men and women, took the vows of truth, *ahimsa*, celibacy, non-stealing, non-possession and control of the palate, and dedicated themselves to the service of the people. Soon after opening the Ashram, Gandhi admitted an untouchable family. This was a revolutionary thing. People were furious. The ashram was run by the gifts of the rich neighbours. They stopped it. But Gandhi remained unperturbed. Just then an unknown person came and gave him a purse, which contained enough money to run the ashram for a year. By that time, the storm had subsided.

Gandhi delivered his first speech in India on the occasion of the opening ceremony of the Banaras Hindu University on February 4, 1916 The audience consisted of many magnets and princes and the Viceroy himself. He began by expressing his deep humiliation and shame at being compelled to address his countrymen in a language that is foreign to him. He said; "His Highness the Maharaja who presided yesterday over our deliberations spoke about the poverty of India. Other speakers laid great stress on it. But what did we witness in the great *pandal* in which the foundation ceremony was performed by the Viceroy (Lord Harding). Certainly a most gorgeous show, an exhibition of jewellery, which made a splendid feast for the eyes of the greatest jeweller who chose to come from Paris. I compare with the richly bedecked noblemen the millions of the poor. And I feel like saying to these noblemen; 'There is no salvation in India unless you strip yourselves of this jewellery and hold it in trust for your countrymen in India'." He added, "Whenever I hear of great palace rising in any great city of India, be it in British India or be it in India which is ruled by our great chiefs, I become jealous at once and I say: 'Oh, it is the money that has come from the agriculturists.'"

"While the Viceroy was going through the streets of Banaras", added Gandhiji, "there were detectives stationed in many places. Is it not better that even Lord Harding should die than live living death?"

Many princes and high officers did not relish Gandhi's speech and left in disgust. Even the Maharaja of Baroda who was presiding over the function left. Then only Gandhi ended his speech saying, "Now that the chairman has also left, I must stop."[6] This address of Banaras Hindu University can rightly be called the beginning of Mahatma Gandhi's public life in India. In 1917, at the request of the poor peasant Raj Kumar Shukla, Gandhi went to Champaran, in Bihar, to inquire into the grievances of the peasants of that district. The British indigo planters compelled them to grow indigo on 15 per cent of their land and part with the whole crop for rent. Thousands of peasants who had been suffering from the curse of this system rushed to narrate their vows and miseries to Gandhi. This annoyed the indigo planters and Government. The police superintendent ordered Gandhi to leave the district. Gandhi refused. This refusal was a new thing and electrified the people. He was summoned to appear in court the next day. Gandhi pleaded guilty. This embarrassed the magistrate who postponed the trial and released him. Later Gandhi was told that the case had been withdrawn and he was free to proceed with his inquiry. The Commissioner even offered to give every possible help. While conducting the inquiry, Gandhi preached the principles of *Satyagraha*. He asked the peasants to free themselves from fear. He called for volunteers to teach the illiterate and ignorant peasants in elementary hygiene. He opened schools for their children. He told the people to be ever conscious about their duties. His ever-increasing popularity was troubling the Government. To turn him out they acceded to his request to set up a committee of inquiry. Gandhi himself was member of that Committee. Its report went in favour of the tenant farmers. And thus was the stain of indigo washed from Champaran. The success of his first experiment in *Satyagraha* in India increased Gandhi's reputation and made him a hero.

Gandhi had hardly finished his work in Champaran when he got an urgent appeal from the textile workers of Ahmedabad. Their disputes with the mill owners were taking a serious turn. Gandhi reached Ahmedabad and satisfied himself that the 'workers' demands were legitimate. The mill owners stand, not to submit the dispute to arbitration, was unreasonable. Gandhi asked the workers to go on strike, only if they took a pledge to remain non-violent. The mill hands took this vow. But after two weeks, their spirits began to sag. Gandhi was afraid that they might break their pledge and resort to violence. At this point, Mahatma Gandhi decided to fast. He declared that he would not touch food until a settlement had been reached. The fast hardly lasted for three days when both parties decided on an arbitration and ended the strike amid general rejoicing.

Soon after this, to the Kaira (Kheda) district of Gujarat called Gandhiji. In early 1918, the crops of Kaira district were washed away by rain. The farmers were left with nothing and could not pay the land revenue. They prayed for the exemption of land tax. But the Government turned a deaf ear to their request. Gandhi advised *Satyagraha* but he wanted "at least one of the workers of Gujarat Sabha to accompany him and devote all his time to the campaign until it was completed." Vallabhbhai offered his services, much to Gandhi's delight. When Gandhi appealed to the people to refrain from paying land revenue, the Government became furious and enforced punitive measures for tax collection. Lands were attached, property confiscated, and cattle were auctioned. This meant too great a hardship for the famine hit people of Kaira. The peasants had faith in their leader and stood firm. Ultimately, the Government had to yield. It agreed that the tax should be collected only from those who could pay it. This was what the Gujarat Sabha had been asking for. So the no tax campaign was called off.

The Viceroy Lord Chelmsford invited Gandhi in 1917 to War Conference convened in Delhi. The Viceroy asked him to enlist the support of Indian leaders for the recruitment campaign. Gandhi then believed that the British Empire was on the whole good, and that since India had benefited by the British connection, it was the duty of every Indian to help the Empire in its hour of need. Gandhi supported the resolution of the War Conference. He also toured Kheda district in the summer of 1918 to enlist men for the army. Gandhi had also hoped that his efforts and country support to the British would be rewarded by granting India "Home Rule".

It may be added here that whereas Gandhi was in favour of unconditional help to Britain to wage the war, Rajaji wanted to tell the Government plainly that they were prepared to "help if the Congress Government would give there the preliminaries."

In pursuance of the Montagu Declaration of August 20, 1917, the British Government made a gesture by removing the racial ban, which excluded Indians from the King's Commission in the army. To this Rajaji said "Tilak had made it clear that it would be idle to ask the intellectual classes to join the army unless a change was made in the political situation." Subsequent events have proved that Gandhi was an idealist and Rajaji was more practical.

The coming of Mahatma Gandhi on the Indian political scene was a revelation not only to the rulers of India then but also to the millions over whom they held sway. He brought a glimpse of hope to the hapless masses sunk in poverty, content with their miserable conditions of life, dragging on their existence from day-to-day in darkness and despair, with no hope of the dawn. Jawaharlal Nehru describes with a realistic touch the prevailing state of affairs:

"And then Gandhi came. He was like a powerful current of fresh air that made us stretch ourselves and take deep breaths, like a beam of light that pierced the darkness and removed the scales from our eyes, like a whirlwind that upset many things but most of all the workings of people's minds... The essence of his teaching was fearlessness and truth and action allied to these, always keeping the welfare of the masses in view. The greatest gift for an individual or a nation, so we had been told in our ancient books, was *abhaya*, fearlessness, not merely bodily courage but the absence of fear from the mind. The dominant impulse in India under British rule was that of fear, pervasive, oppressing, strangling fear... It was against this all-pervading fear that Gandhi's quiet and determined voice was raised: Be not afraid."[7]

Rajaji had risen in Salem from a humble lawyer to the heights of his profession. At the suggestion of friends like Kasturi Ranga Iyengar, who was then Editor of *The Hindu*, he had migrated to Madras to practice in that metropolitan city, famous for its brilliant galaxy of lawyers and brilliant juniors at the bar. Apart from finding a bigger field for his talents at Madras, Rajaji's intention was to take more interest in public life and actively associate him with politics. Gandhiji's first contact with Rajaji is best described in his own words:

"Mahadev Desai very soon formed acquaintance with Rajagopalachari who, from his innate shyness, kept himself constantly in the background. But Mahadev put me on my guard. 'You should cultivate this man' he said to me one day. And so did I. We daily discussed plans of the fight together, but beyond the holding of the public meetings, I could not then think of any other programme."[8]

Gandhiji then had come to Madras to consult with local leaders about his idea of starting civil disobedience against Rowlatt Bill if it was passed into law. Kasturi Ranga Iyengar convened a meeting of Congress stalwarts, which included C. Vijayaraghavachari. While they were discussing the matter, Gandhiji felt he could not solve the problem and confessed to Vijayaraghavachari that it was beyond his capacity. Meanwhile the Rowlatt Act had become law. "While these cogitations were still going on," continues Gandhiji, "news was received that the Rowlatt Bill had been published as an Act. That night I fell asleep while thinking over the question. Towards the small hours of the morning I woke up somewhat earlier than usual. I was still in that twilight condition between sleep and consciousness when suddenly the idea broke upon me—it was as if in a dream. Early in the morning I related the whole story to Rajagopalachari. 'The idea came to me last night in a dream that we should call upon the country to observe a general *hartal*. *Satyagraha* is a process of self-purification, and ours is a sacred fight and it seems to me to be in the fitness of things that it should be commenced with an act of self-purification. Let all the people of India, therefore, suspend their business on that day, and observe the day as one of fasting and prayer. The Mussalmans may not fast for more than one day; so the duration of the fast should be 24 hours. It is very difficult to say whether all the provinces would respond to this appeal of ours or not but I feel fairly certain of Bombay, Madras, Bihar and Sindh. I think we should have every reason to feel satisfied even if all these places observe the *hartal* fittingly.' Rajagopalachari was at once taken up with my suggestion."[9]

It will be seen that some mysterious force so drew Gandhiji and Rajaji to each other at their very first meeting that the Master takes the follower into his confidence and reveals to him what passed in his mind during a vision! This strange coming together of these two remarkable persons augured well for the future of the country. Thereafter, Rajaji became a most devoted disciple of the Mahatma, recognizing in him one who had been charged with a divine mission to lead India to the threshold of freedom. This first instance of Gandhiji confidence in him had a psychological effect on Rajaji's mind. For, from that day he became a most valued colleague and co-worker of the Father of Nation, who sought his counsel whenever he was in doubt about the correctness of the path he was taking. At the same time, Rajaji realised that Gandhiji was a leader far out of the ordinary, gifted with a vision that baffled the most experienced among them, destined to lead the country from subjection to freedom. He did not become a blind follower carried away by the charisma of his personality but he considered the pros and cons of a question and gave his loyal support to Gandhiji only when convinced of its correctness.

Rajaji considered that Gandhiji's worship of truth and non-violence were not mere truth of words and abstinence from pleasures of life. He wanted people to adopt these two great principles of good conduct in their day-to-day life. Rajaji fully appreciated the meaning of *ahimsa* as expounded by Gandhiji who declared: "*Ahimsa* is not for saints only but for all ordinary men in their own life pursuing their common objectives in the world." But the strict adherence to Truth and Non-violence of Gandhiji's concept is no easy matter for it calls for a heroic faith in God and indomitable belief that God will be by us in our adoption and practice of these two great ideals. Truth waited like a vigilant spy on Gandhiji and his inner voice always spoke to him in sure though soft tones. Gandhiji lived and worked on no man's sufferance. He suffered the world but he himself was free. Of no one else, I say with great regard, can this be said."[10] Gandhiji's faith in *ahimsa* was attacked by his carping critics when he directed that the life of a calf at the Sabarmati Ashram, which was in mortal agony, should be put an end to, in order to relieve its suffering. "The Apostle of *Ahimsa*," they cried in one voice, "to let a calf be killed!" Gandhiji asked them not to convert non-violence into idolatry for it would degenerate into violence if blindly followed without reference to time and circumstances. So Rajaji soon became a votary of truth and non-violence and whole-heartedly approved of Gandhiji's plan of non-cooperation for winning Swaraj for India. Gandhiji never promised his followers any prizes if they followed him in thought and action, except the inestimable prize of independence for India. For a nation to obtain non-violent victory against the mightiest empire in the world has known, patience and faith were absolutely necessary, even as their shield and buckler were their unconquerable faith in God and His Grace.

Gandhiji was a politician, no doubt, in his quest for India's *Swaraj* but he had a spiritual mission basing all his work in the Supreme. Had he not declared long ago that he had entered politics in order to spiritualise it? This spiritual basis of Gandhiji's movement for India's freedom attracted Rajaji and made him one of his most trusted lieutenants, loyal to the core, living in the shadow of the leader ever afterwards, fulfilling his ideals in his own life as no other colleague or co-worker of the Mahatma did. Rajaji's reverence for Gandhiji and his rare life of service and sacrifice in the cause of the country grew with the passing of the years and exalted him in Rajaji's estimation to the extent of looking upon him as the Matsya *Avatar* of our age. Many were the splendorous tributes paid to Gandhiji during his Centenary year of 1969 by eminent personalities of the East and the West. Rajaji joined them in offering his garland of flowers to the memory of the Mahatma:

"Others have thought, I myself have thought, the things that Gandhi thought, sometimes even before he put them down for others to read. But the power that enabled him to do what he thought fit came from above. So it is I consider him one who held a commission from God, an *Avatar*. The politics of India may go its own way. Gandhiji now belongs more to the world than to India."[11]

Gandhiji's achievement was not merely freeing his people and his country from foreign bondage. The hero of humanity that he was, his ambition was to bring about harmony between the nations of the worlds especially after seeing the horrors of Hiroshima and the holocaust of two world wars. Gandhiji aimed at the realisation of "One World", the dream of Windell Willkie, and the building up of World Government. Rajaji was aware of this aspect of his Master's efforts and endeavours for world peace and followed in his footsteps. Just as Gandhiji concept of truth and non-violence coloured all the thoughts and actions of Rajaji, this inspiration for working for world peace had its source in Rajaji's adherence to and adoption of his Master's ideals. Rajaji not only dinned into the ears of the stream of the Big Powers to follow nuclear disarmament and give up their lust for dominating the earth on the strength of their stockpile of nuclear arms, he also actively preached the doctrine of unilateral action in order to achieve the noble objective of world peace. To put his ideas into practical shape, Rajaji led a World Peace Mission in 1963, under the auspices of the Gandhi Peace Foundation, to the United States and interviewed President Kennedy and the leading statesmen and thinkers of America. His mission was not without visible results in the minds of the American people. For President Kennedy declared after the interview that Rajaji had a "civilizing influence" on him! The worshippers of power admitted the futility of force and the supremacy of spiritual strength when propounded by this sincere and true follower of Gandhiji. "Gandhiji was not what is usually denoted by the word 'pacifist,' although the pacifists of the world were attracted to and hung round him because of his illustrious career and success with non-violence. He had his own definitions in all matters. He believed in truth and non-violence but he believed equally in resisting evil, in which he included international ambitions. He was not a non-resister, but he loathed shedding the blood of anyone in his resistance except his own. He therefore called himself a non-violent resister, not a pacifist. If we believe in national frontiers and in guarding them, we should not talk the language or think in terms of pacifism but refer rather to the Gandhian dictum. We know what Gandhiji wanted people to do, if they were not ready to sacrifice involved in non-violent resistance. He permitted armed resistance in good causes when there was no reasonable chance of offering non-violent resistance. And armed resistance includes not merely offering force in desperate isolation but obtaining the support and co-operation of helpful friends."[12]

It did not take long for Rajaji to realise the spiritual eminence of the Mahatma, though the latter moved and worked with ordinary men and women from day-to-day, whether in the field of political action or for securing equality. Rajaji considers that Gandhiji raised India to a position of high dignity among the nations because of his moral stature and his insistence on the ideals that form the fine core and foundation of Indian culture. Rajaji reiterated this aspect of the Mahatma's influence on Indian life and thought when he saw a sad decline in the character of the people and their rulers after Independence. What did Gandhiji want of us, he asks and answers the question: "Gandhiji desired that we should lead industrious lives and banish laziness from the land. He wanted us to lead simple lives and treat all our possessions as a trust held for those around us, which is the only non-violent means to banish the disparities of wealth. He wanted everyone of us to banish avarice, hatred and envy out of our hearts and keep means to happiness and the most effective form of true patriotism." Rajaji, in consonance with his following the fine ideals of his Master, appealed to the people of India, especially the elite, to live in such a way as to bring back the moral grandeur that Gandhiji gave us in our days of struggle for *Swaraj* and restore India to her proud position of honour that the nations of the world accorded her during his lifetime.

It was the spiritual caliber of Mahatma Gandhi, more than anything else that made an unforgettable mark on Rajaji's mind. A man who relied on reason and logic to the last degree, Rajaji often battled against the "inner voice of the Mahatma," whose avowed attachment to truth often triumphed over the intellectual powers of his protege. When he differed on a point with him, Rajaji would not easily allow himself to be convinced or conquered by the sheer weight of Gandhiji's moral ascendancy. He would argue with Gandhiji till the last and only when he felt floored by the faith of his Master, who declared that "Faith transcends reason," Rajaji would allow his conscience to accept the opposite view. Otherwise he would agree to differ and maintain his point, though most devoted to his leader in a most personal way. This was exemplified on a number of occasions but the most notable instance of such divergence of views between leader and lieutenant was over the question of Pakistan. Rajaji was in favour of making such a concession in principle to Mr. Mohamed Ali Jinnah, the Muslim Leader, in order to make him give up the idea of demanding a separate state for the Muslims. When his opinion found no support from Gandhiji, there was no alternative for Rajaji but to leave the Congress and for a time to go into the political wilderness. He struck to his guns alone, undaunted by criticism from colleagues in the Congress. But his vision and wisdom proved true in the end when Britain divided India into two separate States enjoying independence after 15 August 1947.

Jawaharlal Nehru always had a deep admiration for the independence of thought and individuality of action of Rajaji. As long ago as 27 January 1928, he recognised the integrity and personal excellence of Rajaji in a letter addressed to him:

"Though I was beginning to detect some differences in view point between you and me, I had no notion whatever of the terrible extent of these differences. The differences between you and me appeared to be so vast and radical that there seems to be no meeting ground between us. I can't conceal from you my grief that I should lose a comrade so valiant, so faithful, so able and so honest as you are and have always been but in serving a cause, comradeship—if dissolution must come in no way effects our personal intimacy. We have long been members of one family and we remain as such in spite of grave political differences."[13]

While they were the most loyal of comrades in the fight for India's freedom, both owing unalloyed and unsurpassed allegiance to their leader, the gap between Rajaji and Jawaharlal Nehru grew after Rajaji left the Central Cabinet and confined himself to the more glorious role of an elder statesman of the nation. He took upon himself the duty of admonishing the government and people for betraying Mahatma Gandhi's ideals and straying from the path of probity in the administration. Before he assumed this unenviable role, Rajaji had been chosen to succeed Lord Mountbatten as the Governor-General of India. All the political parties in India as well in Great Britain acclaimed his selection. Jawaharlal Nehru then heaped unqualified praise on his assumption of the high office. Unveiling his portrait in the Madras Legislative Assembly on 24 June 1948, he also referred to his undisguised divergences of opinion from Gandhiji:

"Rajaji is a person of strong convictions and strong beliefs who has adhered to them. You know he was very much influenced in the early years by Mahatma Gandhi. He accepted basically the approach of Mahatmaji to various problems. Nevertheless, at no time did he become a person who unthinkingly accepted anyone's dictum, not even Mahatmaji's. Rajaji would argue with Mahatmaji, debate with him and would not give in easily. Ultimately, however, he gave in because he thought it best under the circumstances to accept Mahatmaji's judgement. Rajaji's faith in Mahatmaji was tremendous but he never allowed his mind to slip away. Whatever happened, he came to his own conclusion. But if he did give in to somebody, he did so consciously and deliberately, after arguing the matter. Therefore, Rajaji, whether he sometimes disagreed with us or very often agreed with us, brought an extraordinarily keen and analytical mind to bear on every question that came before us. Whether he agreed or disagreed with us, he would like every question to be discussed thoroughly. So we honour tremendously a man like

Mahatma Gandhi for what I call the opposite of possessions, titles etc. Likewise we honour Rajaji for his spirit of individuality, service and sacrifice. Rajaji represents fundamentally the highest type of mind of India.[14]

From the time he began to know Gandhiji and became his devoted follower, Rajaji was struck by his Master's unswerving adherence to the principles he avowed and actively practiced whether in the political field or in his private life. Gandhiji's basic philosophy was that every man is intrinsically good and reasonable and if he is properly approached, he will turn from the path of error to one of rectitude. Gandhiji therefore offered the method of *Satyagraha* or self-suffering in order to convert an opponent to one's own views, thus eschewing violence to achieve that end. "Gandhiji did not accept that only good men will yield to self-suffering, not bad type of men, and that *Satyagraha* was not good against the latter. Gandhiji believed in the Hindu Scripture: "*Isvaras Sarvabhootanam Hriddesha Thishtathi*: 'The Lord dwells in every one's heart.' The assumption that every human being is basically good and that we can and should appeal to that element in him stems from the Hindu philosophy of *Vedanta*."[15]

The personal interest that Gandhiji evinced in his immediate colleagues and co-workers struck Rajaji as marking out the Mahatma as being different from most of the political leaders of the land. Once his activities were restrained by Government throwing him into jail, Gandhiji confined himself to his own thoughts and communing with his Maker. It was because of this spiritual background of his whole like that he could boldly declare that God guided every moment of his life. Even in his letters from prison, Gandhiji was scrupulously non political, more Interested in the private and personal affairs of his lieutenants, curious to know how they were facing rigours of jail life. Rajaji mentions one letter he received from Gandhiji in Bardoli dated February 3,1922.[16]

"I was glad and thankful to get your note. You certainly miss nothing by not having newspapers. And I do envy your spinning wheel and *Ramayana*... You must not lose flesh. I thought, in your company, I was the lowest in weight but you beat me fully by 10 lbs. now. I hope you are allowed plenty of milk. Let me prescribe for you. If you can get milk and plenty of sweet juicy fruit, oranges or grapes, you will get rid of your asthma and gain in weight. Devdas is in Allahabad. Kristodas and Ramdas look after me. You are not the only one to enjoy your solitude. Sundaram is just now in the Ashram undergoing a vow of silence. He has read too much and thought too little. His silence may do him good."

Rajaji was struck by the extraordinary humanism of Mahatma Gandhi and imbibed a great bit of it himself. During his many tours throughout the length and breadth of India, Gandhiji realised how deep was the poverty under which the masses were sunk, how miserable was there existence and how they were at the mercy of landlords and moneylenders whose grip over them seemed to be unbreakable. Gandhiji was horrified at the sight of the skeletons that had greeted him in the famine-stricken areas of the land. His soul rebelled at the existence of the institution of untouchability in Hinduism, which he considered a parasite on its grand and glorious philosophy. On a memorable occasion he had declared, "Untouchability is like a blot on the fair name of Hindu Religion". He determined to remove untouchability root and branch, which was one of the main items in his programme for the regeneration of India. He undertook an all-India tour for getting shrines and temples thrown open to the untouchables, whom he christened Harijans or the Children of God.

Rajaji was so inspired by Gandhiji's gospel of equality for all classes of Hindus that he accompanied Mahatmaji on his long pilgrimage to rid the holy places of Hinduism of their perverse idea that the entry of untouchables would pollute the deity in the shrines. "You must have a poor opinion of God", asserted Gandhiji, "If you think that the entry of Harijans into temples will pollute them instead of God purifying them and the other Hindus of the four castes." Gandhiji made up his mind not to step into any temples that denied access to Harijans for worship therein like other Hindus. The agony rankled in his souls so deeply and he identified himself so much with the cause of the Harijans that he once said with undisguised emotion, "I wish I am born an untouchable in my next birth!" Not only did Gandhiji devote his energies and enthusiasm for the banishment of untouchability from the Hindu fold, he often lived among them. Rajaji in his own time played no mean part in his Master's endeavours to see that justice was done to the lowest in Hindu society.

The semi-nakedness of Mahatma Gandhi's body revealed to the world the spiritual grandeur of his soul before which kings and commanders quailed and which evoked the spontaneous reverence of man all over the world from East to West, equaling the homage paid to the noblest saints in history. Rajaji again and again refers to these sublime aspects of his Master's life, whom he deemed an *Avatar* come down into the world to conquer the hosts of evil by his selfless service of mankind and supreme sacrifice for the cause dear to his soul. In Rajaji's opinion, Mahatma Gandhi was a miracle wrought by God even as Socrates and Buddha and Jesus Christ were miracles come out of His Immortal Mind! Gandhiji's views are good for all time since they are the echoes of the Everlasting and Eternal Spirit. Rajaji quoted from Robert Payne's book on Mahatma Gandhi published during the Centenary Year, a passage that portrays ineffable language the lofty place that Gandhiji holds among the heroes of humanity:

"In some strange and mysterious way, the plain white loin-cloth and shawl of a Hindu peasant looked like the robes of an Emperor on Mohandas Karamchand Gandhi. It was the man himself, of course, who made this so, but who but an Emperor or a saint could conquer the heart of the country and become its conscience? It was the example of his life and his belief in non-violence, the new dignity he gave to the villagers, the simple assertion of human rights. He fought for elemental things, for truth and justice above all, and he was able to give weight and meaning to these words, which were otherwise weightless and meaningless. Future historians will probably regard him as one of those rare men who come to an end of historical epoch and by their very presence announce the beginning of a new dispensation, though they are not themselves permitted to see the Promised Land. He was one with Buddha and the ancient sages."[17]

REFERENCES

1. Ahluwalia, B.K. and Shashi Ahluwalia, *Rajaji and Gandhi* (New Delhi: Allora Publications), 1978, pp. 37-38.
2. *Ibid.*, p. 38.
3. Doke, J.J., *Mahatma Gandhi* (Oxford: Oxford University Press), 1977.
4. Gandhi, Rajmohan, *The Rajaji Story: A Warrior from the South, op. cit.*, pp. 31-32.
5. Sharma, Jai Narain, *Power, Politics & Corruption: A Gandhian Solution* (New Delhi: Deep and Deep), 2004, p. 113.
6. *The Collected Works of Mahatma Gandhi* (New Delhi: Publications Division, Ministry of Information and Broadcasting, Government of India), Vol. XIII, p. 564.
7. Nehru, Jawaharlal, *The Discovery of India* (New Delhi: Sage Publishers), 1977, p. 427.
8. Gandhi, M.K., *An Autobiography or The Story of My Experiments with Truth* (Ahmedabad: Navajivan Publishing House), Vol. II, p. 484.
9. *Ibid.*, pp. 485-487.
10. *Swarajya*, 2 May 1964.
11. *Ibid.*, 4 October 1969.
12. *Ibid.*, 16 February 1963.
13. Nehru, Jawaharlal, *A Bunch of Old Letters*, p. 56.
14. *Swarajya*, 11 November 1967.
15. *Bhavan's Journal*, 4 July 1965.
16. Rajagopalachari, C., *A Jail Diary* (Madras: Rochouse Publishers), 1941.
17. Payne, Robert, *The Life and Death of Mahatma Gandhi* (New Delhi: Sage Publishers), 1977.

3

ON FREEDOM

Rajaji had passed the psalmist's span of three score and ten years when he was last in power as the head of a government. As the Chief Minister of Madras in 1952, he had brought stability to the Congress Party in that State and acted as a bulwark against Communism. At the request of Jawaharlal Nehru, he took over the administration of that State again. But the disease of the communal discords which dominated the minds of other leaders, did not permit him to continue in power for long. He bade goodbye once and for all to controversial politics in the Congress Party and soon become the mouthpiece of the voiceless millions of the land. Thus, having been a fighter for the freedom of his country for over a quarter century, he soon became a fighter for the freedom of his countrymen who had yet to reap the benefits of *Swaraj*.

At this juncture, Rajaji found India at the crossroads—divine destiny after centuries of subjugation awaited her or a doom that men could not contemplate! A decade of freedom had brought her people not all the blessings. The Father of the Nation had led the people to the threshold of liberty. But the fruits of the freedom were as distant as ever from the common man, who groaned under poverty and grovelled in misery. The rich and the industrial classes had grown richer, capitalising on their moneybags and monopolising their influence with officialdom. The poor had grown poorer with soaring prices, high taxation, the horrors of corruption dogging them. There was more want in their homes, more callousness on the part of pitiless officials, more unemployment in the land, than even before. The late Shri D. Sanjivayya had said on one occasion, when he was the President of the Congress, that those who had been paupers during the freedom struggle, were living like princes after India attained Independence.

All this was due to the fact that those enjoying power and authority had betrayed the masses. The only hope of redemption for the country was the replacement of the present administration by leaders, whose ideal was integrity in public and private life, whose ambition was the use of their high authority for the amelioration of the condition of the common folks, whose patriotism was above temptation of any kind and could not be prostituted for perverse ends and personal gain. Therefore, Rajaji took upon himself this noble mission of providing such leadership to the country and gathering around him a band of dedicated colleagues and co-workers, diverting their energies into fruitful channels of service to the people, men who could reclaim the country by their own example and precept from the moral degradation into which it had sunk. This moral degradation gave birth to Swatantra Party, the Swatantra Party, with Rajaji's blessings.

As early as 1921, soon after the start of Non-cooperation Movement, Mahatma Gandhi had warned the nation against seven sins. These were according to him, "Politics without principles, wealth without work, pleasure without conscience, knowledge without character, commerce without morality, science without humanity and worship without sacrifice." Rajaji found infamous lapses from virtue and the good life raising were their ugly heads all over the land after Independence. It was true that Gandhiji had designated Jawaharlal Nehru as his political heir regarding whom he had said with unalloyed pride as if he were his own son:

"Who can excel him on the love of the country... If he has the dash and rashness of a warrior, he has also the prudence of a statesman...He is pure as crystal, he is truthful beyond suspicion. He is a knight sans peur, sans reproche. The nation is safe in his hands."[1]

But, alas! The prophecy of Mahatma Gandhi that Jawaharlal Nehru would make good as an able administrator proved untrue in spite of his integrity. No doubt, he had many excellences and virtues in his fascinating personality but he failed to make the mark as a statesman. His head was too much in the clouds; feet were firmly planted on earth. He loved flattery and sycophants soon gathered around him, burning incense at the shrine of his self-pride. An unbounded admiration for Soviet Russia made him copy its industrial methods, stifling the individuality of man in his enthusiasm for rapid progress, and yearning to raise his people to the level of Western standards of life. This led him to embark on policies and programmes that were not in harmony with the ancient culture and civilisation of India that even today influences the common folks. India lost the exalted position to which the Father of the Nation had raised her during the freedom struggle. The doctrine of Non-alignment, enunciated by Jawaharlal Nehru followed with enthusiasm by him in international affairs, soon became a negative approach. The passing away of Jawaharlal Nehru on 27 May 1964 left no dominating personality in the Congress, capable of coping with the situation. "Sri Jawaharlal Nehru urged by patriotic impulses and early indoctrination," said Rajaji, "committed the blunder of taking India out of the path; all he could was to transfer our ideology into that of Soviet Russia. This was the fatal step that has brought us to the present position, out of which it requires not only wisdom but indomitable courage to save India. Social justice and removal of disparities of opportunity and equitably distributed welfare are great and worthy ends. But the fatal mistake was the plan to achieve this by the short-cut of heavy borrowing and central planning and a permit-licence regime, which has brought in its wake all that make Sri Prakasa lament so bitterly."[2]

It saddened Rajaji that the highest in the land were not imbued with a sense of honour or a spirit of honesty in their dealing. When the elite of society were even glorifying the principle that the end justifies the means, how could the lower officials be expected not to stoop in order to make both ends meet? Between the devil and the deep sea, the voiceless millions of the land were ground down so heartlessly by the government that their groans could hardly be heard. As long ago as 1922 when he was in jail and *Swaraj* was yet a dream, Rajaji had uttered and recoded these pregnant and wise words in his Jail Diary, which are applicable fully today: "Freedom is often attained by sacrifice of successive generations and we should therefore be prepared to lay at the altar of the country what we could give or suffer without hungering for immediate fruit—even if we do not feel we are able to reach the *Gita's* ideal of duty without concern for results altogether. Think where we would have been if our father and grandfathers had made definite sacrifices for freedom's sake. Would we not have carried forward then the battle for freedom with greater faith and vigour? By our sacrifices we have at least made the history of India in this generation an honourable chapter, a relief from the continuous story of surrender, indifference and dishonour. Even this is an inheritance for our children, so let us not lose faith."[3]

An ardent believer in democracy and the democratic way of life, Rajaji opposed the idea of a brute majority oppressing the people and suppressing their sacred rights under the Constitution, because of its voting strength in Parliament and State Legislatures. A democratic government must have faith in the loyalty of the people, unperturbed by noisy demonstrations of popular excitement. The confidence of the people in the rightness of its policies is the key to success in a democratic government while in its own turn; it must have faith that it is working for the welfare of the masses. Further, no democratic government should consider all opposition to its programme as hostile. After all, democracy goes from strength to strength when there is a healthy Opposition acting as a check on its activities, which are not in the interests of the nation. "Nations and governments are truly democratic only if they can preserve and abide by their democratic constitutions, even when assailed by difficulties and temptations, and when the rule of law and the fundamental rights may be felt to be a nuisance by an impatient government with a big majority behind it in Parliament which will vote for whatever the government wants, including amendment of the Constitution itself."[4]

Rajaji bewailed the lowering of the standards of purity in the administration and the fall in the probity of its personnel after Independence. Mahatma Gandhi had raised India to a position of honour among the nations of the world. His ideals of truth and non-violence had been given the go-by by the government, which seemed to consider them an anachronism in the sixties and seventies in

India! The rulers and the ruled were blindly marching towards a precipice in their moral degeneration, forgetting the noble ideals of *Satyaam* and *Dharmam*. Rajaji felt that the people had betrayed Gandhiji's faith in them. "All things that Gandhiji wanted to be done are forgotten and the very opposite of what he wanted are being done. He desired to establish a society in which the people had had the largest mount of freedom and the government took the minimum power for itself. The state of things now, as everyone knows, is one of the rapidly shrinking freedoms for the people and increasing power for the State. Gandhiji was against controls with which unfortunately we are now so familiar. The permit-licence is not disappearing but has come to invade our kitchens. Gandhiji wanted taxation to be low and simple in form but we have now crushing taxation, and new and yet newer imposts and levies. He wanted the State to be decentralised. But we have instead, more and more centralisation. The decentralisation we are asked to accept as a great achievement, viz., Panchayat Raj, turns out to be a gigantic fraud to get the Congress Party's work done at State expense by State organisations, by the provincial ministries and panchayats. It is not decentralisation but the octopus of the party spreading itself to live on State taxes and State power instead of its own popularity. Gandhiji wanted character, courage and integrity to be strengthened. But everything done by the Congress Party, using State power tends to destroy courage and to extinguish integrity. Gandhiji sought to spiritualise politics. His very entry into politics he justified on that ground. But we see now, as a result of Congress rule for seventeen years, materialism is spreading and replacing the traditions and ideals of our nation. We see the people, under the pressure of the Government's exhibitionist plans and consequent unbearable taxation, growing more dishonest and unmanly. The elections, the expensiveness of which has become a vested interest of the rich Congress Party with its pistol-control over all the purses of the businessmen of the country, have become more and more bribe-riddled so much so that the people are losing faith in democracy itself. This government must be changed if we do not wish utterly to dishonour the Father of the Nations."[5]

Rajaji made this strong indictment against the Congress Party and its administration during Jawaharlal Nehru's lifetime. But it seemed to have no effect on Nehru or his successors. "The dogs bark but the caravan passes on!" was the comment of Sir Samuel Hoare as Secretary of State for India during the dark days of the civil disobedience movement in the thirties of this century. The Congress leaders seemed to echo his words, keen on perpetuating themselves in power at the cost of moral principles and insidiously converting a democracy into a dictatorship. Rajaji had always raised the voice of dissent against such degradation of the rulers, as the mouthpiece of the toiling millions of the land and as one who had felt it his duty to continue the mission of Mahatma Gandhi and the inculcation of his high principles but it all seemed to no purpose.

Among the objectives, which Jawaharlal Nehru set out and to achieve during his premiership, he gave a high priority to socialism and the building up of "a socialistic pattern of society." He had visited Russia on the occasion of the Tenth Anniversary of the Soviet Republic in 1928 and had been greatly impressed by its achievements. He had it in his mind that once India became independent, she should follow Russian methods for the regeneration of his own people. This ideal had gained strength during the succeeding two decades and when he became Prime Minister, he attempted to copy the plans and purpose of Soviet politicians in the industrialisation of India and to bring it economically on a level with Western countries like the USA, Great Britain and France.

Socialism was, no doubt, an ideal worthy of implementation in bridging the gulf between the haves and the have-nots. To that extent, Rajaji was in agreement with Jawaharlal Nehru, but he differed from latter with regard to the methods of its fulfillment. He opposed Jawaharlal Nehru on the ground that it was against India's *dharma* to rob Peter and pay Paul. The rule of law, known as *dharma*, had always been observed in India from ancient times both by the rulers and the ruled. Whoever violated the strict injunctions of *dharma* has to pay the penalty for its transgression sooner or later, in this birth or in a future birth. It is on this principle of morality that the right to property must be guaranteed to the citizens by the State. "The free way of life as against life under totalitarian rule," declared Rajaji, "is not possible if the properties of citizens, earned or inherited, are not guaranteed against expropriation by the State without just cause open to review by an independent judiciary. It was not for nothing that Jawaharlal Nehru and others had guaranteed this and had inscribed in the Constitution. The holding of what you have earned or inherited is essential for free life, for life different from that of serfs. Vague assurances in electoral manifestos or platform speeches are not enough. What the citizen requires is protection under the Law, which the Constitution gave until now. Under Roman Law, slaves owned what was known a peculiam. Citizens in India want something more than peculiam."[6]

"The evolution of indefeasible citizen's rights and of parliamentary government, out of one man's rule," says Rajaji, "is as interesting as the evolution of man which Darwin had expounded in his Origin of Species. The one is biological science and the other is the history of civilisation. Both were accomplished by struggle and violence. But what is now evolving in India, without violence or struggle, is the remarkable evolution of totalitarian unitary rule out of a Constitution, which sought to guarantee indefeasible rights and a federal type of parliamentary government, the process being marked by complacence and a selfish indifference among the elite in spite of a high degree of intellectual awareness."[7] Rajaji, however, did not feel discouraged at this transformation. When he started his movement for India's Independence, he asserted that this feeling of despair and dejection at this hapless state of affairs could be overcome by an overwhelming faith in us. He instilled that courage into the minds of the masses. More than anything else, we must stick to truth and purity in thought and action.

It is the duty of government to govern justly in the interest of the people, even as it is the duty of the opposition in Parliament or the legislatures of the land to offer fair and honest criticism of the measures of the party in power, its political objectives and programmes. To this extent, both Government and the Opposition should fufill their *dharma* for the nation's good. A famous slogan of Jawaharlal Nehru when he embarked on the industrialisation of India was "Produce or Perish!" But he forgot the fact that production could not thrive under bureaucratic management or official ownership. When a person imbued with the idea of profit takes a personal interest, there is bound to be an increase in production. Rajaji was against Statism or State Capitalism of any kind in order to foster industry and the economic progress of India. "Adopting the methods of Capitalism," he says, "and grafting them on Statism will do not do the trick. This is the crucial point against Statism, State Capitalism, Communism and other allied 'isms'...The change in methods is a virtual confession of failure of the Communist attempt to go against nature."[8] A harmony can be achieved, says Rajaji, between production that leads to profit and the prosperity of the community by adopting Gandhiji's theory of trusteeship of wealth for the commonweal.

The proper and effective functioning of government in a democracy depends on the organised will of the people, which is made known by elections held from time-to-time. A number of parties with their own political programmes go to the polls and the results of the votes cast by the qualified citizens determine which party shall be entrusted with power. Normally, the majority party has the right to form the Cabinet and choose its own ministers, although coalition governments have not been unknown. Rajaji concedes the necessity for the existence of the State but he will not admit that the power of the State should predominate over the interests of the individual. The authority of the State must be kept within certain limits if the freedom of the citizens is not to be jeopardised; in other words, the individual citizen is supreme, not the state. Not only should the citizen have a sense of freedom, but should also be seen actually enjoying that freedom in an effective manner. Putting in a different language, the citizen must be conscious of his freedom and must have the complete feeling of being able to enjoy it.

The consequence is that if the state is considered to be superior to the individual citizen as in totalitarian regimes, the omnipotence of the State works havoc with the freedom of the individual and reduces to a minimum all chances of his enjoying happiness or harmony of life. "It is the citizen that is the reality," declares Rajaji, "and the tangible and unalterable fact. His freedom must be taken for granted for all mechanism, which is good or bad, being made in parliamentary democracies out of a majority of votes and many other circumstances, not altogether perfect or even good. The frontiers of this mechanism created by citizens for their own good must be strictly and carefully limited. The frontiers of the citizen's freedom must be as wide as possible."[9]

Rajaji reminded people that Mahatma Gandhi was in favour of "minimum government" and his ideal was adopted by the framers of the Constitution, who were among the wisest and the ablest lawyers and leaders of the land, after more than two years of labour. A most important part of the Constitution is that dealing with the fundamental rights of citizens in a free India, which has been inscribed in the forefront of other privileges of citizens. The policy and programme of the Congress Party, which had been all along in power, had been to erode these basic and essential rights, jeopardising the substance of freedom. Rajaji reminded the people that eternal vigilance was the price of liberty and it must be fought for and preserved, sparing no effort or considering no energy too precious to be wasted. "The call of freedom is loud and insistent. The future depends on the response. When the battle for freedom was fought against Britain, we called for courage and the taking of risk and the response was such that it brought success. In the battle for the freedom of the citizen, which is the true substance of the national freedom, there is the same call for courage and the taking of all the risks involved, and may it be that the response will be equally good."[10]

The system of permits and licenses by the grant of which the Congress government ostensibly desired to prevent the concentration of commerce and industry in the hands of a small number of businessmen, came in for severe condemnation at Rajaji's hands. He considered it a means by which the money-power of monopolists in trade was at the mercy of the authorities, who doled out their favours only to those who contributed to the funds of the Congress Party and in return, helped them to corner a major portion of the country's industry and merchandise. He was for a free economy and fair competition whereby the best products could be manufactured at minimum cost to the exchequer. So he engaged himself in opposing the Congress regime, which he held as being responsible for the above evils in the governance of the country. When he joined the Non-cooperation Movement in his middle age under the leadership of Mahatma Gandhi, his ambition was neither acquisition of fame nor power through politics. But the time had come, he declared, when he should lead a revolt against the Congress government. "I came to the conclusion five years ago—and I hold it was a correct conclusion—that a movement for real freedom, as important and as serious as the movement for Independence from British rule, had not to be vigorously conducted against this misconceived progress towards what will end in the suppression of individual liberty and the development of the State into a true Leviathan."[11] He castigated in no undeserved language the conversion of what was intended to be democracy by the founding fathers of the Constitution into a monolithic State of the type of Soviet Russia with all the dark and desolate features of unfeigned totalitarianism visible on its ugly face!

Rajaji was aware how difficult it was for any Opposition to grow up in strength in order to wean the government from its activities, ostensibly for ushering in Socialism into India. But a bold attempt must be made, asserted Rajaji, in this direction if the country was to be saved from the horrors of Communism. "I do not like dictatorships," confessed Jawaharlal Nehru to the French Journalist Tibor Mende who recorded a series of talks he had with the Indian Prime Minister at the height of his fame. "I do not like authoritarian regime. I do not like techniques adopted by the Communist Party. That again means that I would not like to have it in my country."[12] The hatred for authoritarianism expressed by him does not seem to be shared by his daughter in Rajaji's opinion. He wanted the democratic elements in the country to stand up to the Congress government and its policies, which were not in the interests of the nation at large. He was optimistic that a reasoned and united Opposition was sure to make headway if an honest attempt was made to expose the ill-advised policies of the government.

Allied to his strong disapproval of the devious methods of the Congress government to direct the commerce of the country into conduits of fruitful source for its own party funds was Rajaji's attack on the taking over by the State of trade and the embarking on business ventures. "I have sometimes used strong language," he said, "but what language can be too strong for condemning the baneful system of permits and licenses and the hold that a party in office thereby gets over the entire economic life of the nation, which exploits for strengthening and perpetuating its hold while devoutly hoping it may also help the nation. 'Permit-License Raj' has been condemned by everyone now. People have been seen what it has produced by way of corruption, general moral deterioration and building up of monopolies for a chosen few. The permit-licence-quota system stands disgraced along with the Congress Party that swore by it and exploited it."[13] "The way of free enterprise," said Rajaji, "is the best road to national prosperity for people who desire to live under democratic rules as distinguished from authoritarianism."

Another aspect of the activity of the Congress government towards depriving citizens of their right to free enterprise, guaranteed by the Constitution, was the nationalisation of banks. The prominent banks that have sprung up during the present century have provided a good incentive to individuals in banking habit as in the West and, further, they have been the means of helping businessmen and entrepreneurs. The Congress government nationalised the most prosperous banks which had a large capital and huge deposits with them from customers. Rajaji condemned the proposal when the authorities first mooted it. In his opinion the taking over of the banks was against the interests of the people at large, the shareholders and depositors. All the banks have to follow faithfully certain procedures and any failure to do so will land them in financial ruin. Further, they have to fulfil strictly the laws connected with banking institutions and dare not do anything underhand. Banks are the source of funds for trade, commerce and industry to thrive. Under such circumstances, Rajaji considered that there were no substantial grounds for nationalisations of banks. "It is clear to me that what is aimed at by pro-nationalisationists is not nationalisation but Statism pure and simple. They want the government to have total power over the people's funds and the opportunity to decide to whom the funds should be given. The government does not mind if instead of profit there is loss owing to bureaucratic official management. This attitude is as ruinous as it is senseless. It is a defeatist attitude that Mr. Chavan, the Finance Minister, takes that the nationalisation of banks is inevitable. He might as well say that Communism is inevitable, and we should prepare ourselves for it. There is nothing inevitable if the thing is against the interests of the people. The monopolist transfer of all the banks to the government management is certainly against the interests of the people and it must be opposed with courage and conviction."[14]

The Congress Party, according to Rajaji, invented the slogan of "Socialism" in order to capture the votes of an illiterate and an ill-informed electorate about the plans and purposes motivating that party. "Statism is not the road to socialism. It is the way to authoritarianism at the top and serfdom at the bottom."[15] Rajaji condemned this kind of socialism indulged in by the Congress Party and observed that the journey to socialistic goals lay through adding to the national wealth by offering incentives to workers to increase national production.

Rajaji was against the type of "Socialism" that the Congress government was aiming at and the insidious farce that they intended to play on the people by proposing to prepare the country for "a socialistic pattern of society". Rajaji saw through the mask of mischief lurking behind the grandiose scheme, which was nothing else but another brand of Communism. He strove with his timeless vision and tireless vitality to see that India did not become a satellite of Soviet Russia. When Rajaji became the Chief Minister of Madras State for the second time in 1952 he had declared that he was Public Enemy Number One of Communism. "Socialism aims at reducing the disparities of earnings among sections of the people. This cannot be achieved by mere words and promises," emphasised Rajaji, "but only through policies that help a large expansion of jobs carrying just wages and salaries according to ones equipment and talent. Any policy which extinguishes freedom, which is the best and most desirable part of democracy, is not socialism but Communism. To tax people heavily and borrow from abroad, to invest it all in losing concerns, to depend on Communist votes in order to remain in power, neither any of these nor all of them put together make up socialism."[16]

A man of high moral principles and a lawyer of no mean repute and caliber, Rajaji was shocked to witness the manner in which the Congress government treated the Indian Princes in the matter of payment of their privy purses. Sardar Patel, the architect of India's unity, had entered into solemn treaties with the erstwhile rulers of the Indian States by which the Government of India had guaranteed the payment of the privy purses to the princes in consideration of their complete and unqualified accession to the Indian Union and giving up, once and for all, their rights of sovereignty over their territories and other privileges, which they had enjoyed under the British regime for generations. The Congress government, in its bid to implement their new-fangled socialist policies decided, to do away with their agreements soon after Indira Gandhi became Prime Minister. Under her guidance and leadership they committed to the breach of the sacred contracts entered into with the princes, not to speak of their gross violation of the plighted word given to them by Sardar Patel, the then Home Minister of the Government of India. The hero of the Bardoli Satyagraha Movement in 1930, was not only a strong man in tackling serious and delicate situations, his strength also lay in his strict adherence to principles of honour and honesty and the ideals of Mahatma Gandhi. Rajaji was against the deprivation of the privy purses of the princes by amending the Constitution. He felt it was a breach of moral faith with them to go back on the pledges solemnly given by Sardar Patel on the behalf of the Government of India. Rajaji's sense of justice and fair play in the Congress government was rudely shaken when they sought to save a few crores of rupees in this ignoble manner, just to appease a rabid group of members of their own party. He felt that the government had fallen from the ideals of the Father of the Nation in their treatment of the princes, contrary to all canons of law and international convention. "To repudiate one's own obligations after getting full performance from the other side," said Rajaji, "is disgraceful and unwise.... The immorality of the attempt to deceive the rulers who became Indian citizens and the voters in the Union of India, on the basis of the assurances given and after the full performance by the rulers of their part of the contract, is something that cannot be described except by the strongest words of condemnation."[17]

The Congress government, observed Rajaji, seemed to believe that all criticism of its policies and actions was an obstruction blocking the smooth tenor of its administration. This was a mistaken idea on the part of the authorities, for they seemed to forget that the criticism offered by the Opposition parties was healthy assistance, which was absolutely necessary for the growth of true democracy. "Democracy ceases to exist if there is no provision of either sort for free criticism of the policies of the government. Windows to let in air are wanted, not glass mirrors which reflect what is inside but which shut out ventilation. Parliament and the State legislatures are there, but Parliaments in which opposition is over ruled by brute majorities are not enough and are no substitute for public criticism."[18]

In the best governed democracies of the West like Great Britain and the United States of America, the Opposition has a prominent and important role to play in pointing out the faults of the administration and issuing a warning to the authorities not to embark on a policy which jeopardises the interests of the nation. An Opposition is a necessary nuisance and cannot be done away with if a democratic way of life is to be followed. "Without a vigorous Opposition and full play given to it, the Congress Party will become a mutual benefit association. If the claim that India is governed under the parliamentary system of democracy is to be sustained, it must be clearly recognised that criticism must always accompany government proposal and performances. Criticism from opposition parties and members of the public is not just an irritation, which dogs the process of government. It is the fulfilment of a duty on the part of free citizens and an essential of a parliamentary government. The critic is a colleague, not a tolerated seditionist."[19]

Politics played an important part, no doubt, in the life of Rajaji but he always insisted, like his Master, that ethical principles should be the basis of men's thoughts and actions, in the administration of the country. Having thrown himself into the whirlpool willingly and voluntarily, he always kept his head above the turgid waters churned up from the bottom. He called upon his fellow countrymen to fulfil the great destiny of India in the future by following her ancient *dharma* and by implementing the ideals of Mahatma Gandhi, which brought India her freedom, and her fame, as a great nation for rare integrity and wisdom. He warned the rulers in power to give up their wrong views of statecraft, merely aping some of the ambitions of western nations who worship prosperity at the expense of eternal principles. He called upon his countrymen not to hug to their bosoms a materialistic philosophy of life but to keep their eyes ever fixed on the stars in the firmament above. The soul of the nation must be saved, he declared, and no cost should be considered too great or heavy. "Should we not defend freedom?" he asked in an eloquent appeal, "Should we not defend democracy, should we not defend *dharma*? We must not allow India to be debased, mistaking mass selfishness for patriotism. We must love India and must learn why we should love India. We must fight Statism, we must fight one-party rule, which has devised sufficient self-generating power to perpetuate itself, and we must stem the tide of greed and corruption. We must again bring into being a cadre of sternly honest officials to administer our affairs as was done till recently. It is sacred duty, if there be any purpose in life and if we are not just spinning tops. Spinning till death overtakes us and we go down to be re-absorbed into dead matter, fools mourning round the body that once was alive. When the morality of the nation and its elite is being undermined and threatened with destruction, there is no question of alternative or surrender to superior force. We must fight and protect the soul of the nation from being over-whelmed and destroyed. It is the duty of each citizen to resist it to the utmost without waiting for others or counting the cost. For, if the nation's morality is lost, there will be nothing left thereafter to save."[20]

REFERENCES

1. *Young India*, 3 October 1929.
2. *Swarajya*, 18 December 1965.
3. Rajagopalachari, C., *Jail Diary, op. cit.*
4. *Swarajya*, 18 December 1965.
5. *Ibid.*, 17 October 1963.
6. *Ibid.*, 29 January 1972.
7. *The Hindustan Times*, 26 January 1964.
8. *Swarajya*, 16 January 1965.
9. *The Illustrated Weekly of India*, 14 August 1960.
10. *Ibid.*
11. *Swarajya*, 4 July 1965.

12. Mende, Tibor, *Consersations with Mr. Nehru*, p. 98.
13. *Swarajya*, 1 July 1967.
14. *Ibid.*, 19 July 1969.
15. *Ibid.*, 3 January 1970.
16. *Ibid.*, 19 December 1970.
17. *Ibid.*, 26 December 1970.
18. *Ibid.*, 11 December 1965.
19. *Ibid.*, 8 January 1966.
20. *Ibid.*, 5 August 1961.

4

ON DEMOCRACY

Once Rajaji was out of office, he devoted all his time for the betterment of his countrymen. He raised his voice against those policies of the Congress Government, which he felt were not in keeping with the Gandhian principles. The last period of his life is generally known as the period of "*Satyam eva Jayate*" during which he fought for the freedom of his countrymen in free India. Nehru was more inclined to Socialism. Rajaji did not like Nehru's concept of socialistic pattern of society. According to Rajaji, it was a mockery of democracy. He reminded the people that Gandhi was in favour of minimum Government. Gandhi's definition of democracy was, "A nation that runs its affairs smoothly and efficiently without much state interference is truly democratic, where such a condition is absent and the form of government is democratic only by name." This was Rajaji's concept of democracy also.

It is only in a real democracy that people can enjoy their existence. The light of liberty is lost and the lives of the people become desolate if a democratic way of life cannot be ensured to them. In a democracy the talents and energy of the people are given in a spirit of service and sacrifice for the building of the nation and the common weal. Whereas under a totalitarian regime the labour of the people has to be whipped out of them under force at the whim and fancy of dictators, who have no faith in the divine origin of man or his divine destiny. "Freedom is Kaamadhenu.[1] It is better that citizens serve the nation out of free will and in a free way than be drilled, brainwashed and regimented and work as automatons of a political party in power."[2]

Rajaji told the people that the price of liberty was eternal vigilance and should be jealously guarded. He wrote in the *Illustrated Weekly* that the "call of freedom is loud and insistent. The future depends on the response. When the battle for freedom was fought against Britain, we called for courage and the taking of risk and the response was such as brought success. In the battle for the freedom of the citizen, which is the true substance of the national freedom, there is the same call for courage and the taking of all the risk involved, and may it be that the response will be equally good."[3]

Rajaji decried the Government policy of heavy taxations, controls, licences, permit. He said that the Congress Government had betrayed Gandhiji's faith in them. Rajaji used to pour out his mind through the columns of *Swarajya,* which had become his mouthpiece. He wrote :

"All things that Gandhiji wanted to be done are forgotten and the very opposite of what he wanted are being done. He desired to establish a society in which the people had the largest amount of freedom and the Government took minimum power for itself. The state of things now, as everyone knows, is one of the rapidly shrinking freedoms for the people and increasing power for the State." Gandhiji was against controls with which unfortunately we are now so familiar. The permit license raj is not disappearing but has come to invade our kitchens. Gandhiji wanted taxation to be low and simple in form but we have now crushing taxation, and new and yet newer imports and levies. He wanted the State to be decentralised. But we have, instead, more and more centralisation. The decentralisation we are asked to accept as a great achievement, viz. Panchyat Raj turns out to be a gigantic fraud to get the Congress Party's work done at State expense by State organisations by the provincial ministries and panchayats. It is not decentralisation but the octopus of the party spreading itself to live on State taxes and State power instead of its own popularity. Gandhiji wanted character, courage and integrity to be strengthened. But everything done by the Congress Party using State power, tends to destroy courage and to extinguish integrity. Gandhiji sought to spiritualise politics. He justified very entry into politics, on that ground. "But we see now, as a result of Congress rule for seventeen years, materialism is spreading and replacing the traditions and ideals of our nation. We see the people, under the pressure of the Government's exhibitionist plans and consequent unbearable taxation, growing more dishonest and unmanly. The elections, the expensiveness of which has become a vested interest of the rich Congress Party with its pistol control over all the purses of the businessmen of the country, have become more and more bribe-riddled so much so that the people are losing faith in democracy itself. This Government must be changed if we do not wish utterly to dishonour the Father of the Nation."[4]

Rajaji raised his voice when Congress Government wanted to erode the Fundamental Rights by passing the 24th Amendment Bill. He wrote in the *Swarajya*:

"A massive majority of MPs in New Delhi has now resolved to destroy the Fundamental Rights, which our freedom founders, had thought was necessary for the welfare and freedom of our people. As *The Hindu* in the course of its leading editorial of 6 August 1971 said: "The sweeping powers conferred on Parliament by the 24th Amendment to tamper with all the rights, including the right to freedom of speech, of freedom of association and of religion, are such that it is not surprising that some leaders of minorities and even some committed socialists have been alarmed over the grim possibilities."[5]

Rajaji reminded the people that individual freedom is something, which cannot be granted by the government, but it is inherent in the people. In the *Swarajya* he wrote:

"Freedom rests not on Constitutions but on the will of the people to be free. Freedom endures only in the measure and, only so long, as this will last. Liberty is in the hearts of men and women. When it is not there, burning and alive, no Constitution, no law, no court can save it, as an American Judge, put it."[6]

The Constitution of U.S.S.R. is on the face of it a fine document full of promises that point the way to paradise for its people. It enumerates more rights and privileges to them than the American Constitution or the unwritten Constitution of England to its citizens. But what do we see in the actual working of the basic law of the Soviet Russia? The people suffer under the effects of serfdom, they have neither the right of free speech nor freedom of movement anywhere in their country except with the permission of the authorities; when they are shadowed by their ever-vigilant secret police, they dare not oppose the oligarchy at the top of their monolithic State! This is communist doctrine in practice, with no Opposition to criticise, much less condemn the rulers. This is where democracy differs from Communism.

A strong Opposition is necessary *sine qua non* for the successful working of democracy and guarantee of good government, which is unthinkable in a totalitarian State. All those who are suspected of being against the policy of a communist regime are ruthlessly liquidated. Rajaji reminds us that freedom is the fruit of sacrifice made by many generations of men and women in all countries. From the time of Magna Carta, the British people have fought against autocracy and tyranny, even sending one of their reigning sovereigns to the block to be beheaded. In France, till the time of Louis XVI and Marie Antoinette, the people were opposed by their kings and noblemen until they rose in revolt and won their freedom as result of the French Revolution of 1789. The Colonists of America had to rise in the rebellion against the rule of the English kings under the magnificent leadership of George Washington and became independent in 1776. India under the unique leadership of Mahatma Gandhi compelled the British to concede independence to the subjects of their Eastern Empire on August 15, 1947.

For the growth of a true democracy, criticism offered by the Opposition parties was absolutely necessary. Rajaji said: "Democracy ceases to exist if there is no provision of either sort for free criticism of the policies of the government. Windows to let in air are wanted, not glass mirrors which reflect what is inside but which shut out ventilation. Parliament and State legislatures are there, but Parliaments in which opposition is over ruled by brute majorities are not enough and are no substitute for public criticism."

The basic difference between democracy and a totalitarian regime is that democracy depends for its successful working on the will and voluntary co-operation of the people while communism cares but little for the opinions or wishes of those over whom it holds sway. Democracy has no room for violence in its make-up while Communism is based on naked and unabashed violence from top to bottom and bends people to its will by the sheer brutality it practices on those who dare oppose it. In Communism, not only are the bodies of men and women bedeviled with brutality of every kind but also even their minds are moulded into submission by a soulless regime of terror and intimidation. "We do not want to dumb and unmanned population. But we do want democracy, which has no room for violence… Democracies may err but the way to correct errors in democracies is limited by the rule against violence, even if such limitation results in the long run continuance of error demanding patience such as is not easy. It is here that the Vedic motto on our national crest speaks in loud tones, *Satyameva Jayate*. Do not rush to violence in despair and double then wrong but meet it by patience and sacrifice behind powerful argument. Peaceful resistance to authority always involves sacrifices."[7]

With the aim of providing a strong opposition, Rajaji founded the Swatantra Party. He wrote in the *Swarajya* about the aims and objects of his new party. To quote him: "The Swatantra Party stands for the protection of the individual citizen against the increasing trespasses of the State. It is an answer to the challenge of the so-called socialism of the Congress Party. It is founded on the conviction that social justice and welfare can be attained through the fostering of individual interest and individual enterprise in all fields better than through State ownership and government control. It is based on the truth that bureaucratic management leads to loss of incentive and waste of resources. When the State trespasses beyond what is legitimately within its province, it just hands over the management from those who are interested in frugal and efficient management to the bureaucracy which is untrained and uninterested except in its own survival.

"The Swatantra Party is founded on the claim that individual citizens should be free to hold their property and carry on their professions freely and through binding mutual agreements among themselves, and that the State should assist and encourage in every possible way the individual in this freedom, but not seek to replace him.

"The new party seeks to oppose the trend of the ruling Congress Party to adopt the ways and ideals of the communists in its eagerness to prevent the communists from going forward. The Swatantra Party believes that going over to the enemy is not defence, but surrender.

"The Swatantra Party, apart from the ideology here explained, hopes to furnish a real Opposition to the Congress Party so that parliamentary democracy may be properly balanced. The absence of a true Opposition has led to the rapid deterioration of democracy into a kind of totalitarianism. Voices have been heard from all quarters calling for a strong Opposition and the new party is supplying a felt want. This party of freedom is further making a novel experiment in restricting disciplinary control over party members to essential issues, giving freedom in all other matters to vote according to individual opinion. This is not mere strategy to 'net in' discordant, miscellaneous elements as at first might appear. It is really an answer to the constantly expressed sense of dissatisfaction with party rigidity, and to the complaint that it often amounts to suppression of opinion and rule by a minority in the name of a majority. A majority in the ruling caucus can always, under present conditions, impose its views on all and every issue in the Parliament of the nation. The Swatantra Party intends to initiate a departure from the usual practice of political parties and, true to its name, give Swatantra or freedom to its members to vote according to their own convictions (and conscience) on all but the party's fundamentals so that the decisions of Parliament may, on those issues, truly reflect the prevailing opinion, and not be just a replica of the majority opinion of the ruling party or the fads of the ruling clique."[8]

Once Nehru complained that Rajaji was angry. Rajaji replied: "Mr. Jawaharlal Nehru referred to me with respect and affection and expressed his wonder why I speak in so much anger. Let it be called anger, but can he not guess why I am angry? Does he not see how people, who were on the whole honest and God-fearing have now been everywhere made by the policies enforced by Government into a greedy and dishonest people? Does he not realise that a fine cadre of officials with a keen sense of justice, honesty and public service have now been made into spineless flatterers and partisans? I see fear enveloping everywhere like a poisonous fog on all the significant sections of the people. I see God disappearing and mammon occupying his place in the hearts of men. I see traditional values of life centring on compassion steadily being starved out by an all devouring State.

"I can't help being sad. I am not entirely hopeless: so my grief looks like anger.

"The Swatantra Party may not be able to cure the harm already done, but much can be done by way of preventing further mischief."

The Prime Minister said with reference to the Swatantra Party: "Mr. Rajagopalachari has got a medley of companions in the Swatantra Party and it is difficult to know what their stand is. But broadly speaking, he stands for the perpetuation of the traditional system, leaving the industrialists to go ahead and do what they like. Ultimately they will change the system but on a different plane than I would like it to be. But I don't see how by means of that traditional system you can ever get rid of the problem of poverty. Therefore, I cannot imagine how any person can base any approach on merely perpetuating the traditional structure.

To this Rajaji replied, "If I have a medley of companions" in the Swatantra Party, I believe it is not only natural but also fair. Our nation is a medley. No one can deny this or ignore the fact. It is not a good thing that this big country and this large nation should be governed by anything that approaches the homogeneity of a clique. I take the reproach of the new Party being a medley as a compliment. It will not do for a nation or a country such as ours, which will not fit into any framework built on the models of British political history to trust in the uncontaminated homogeneity of a ruling party.

"Mr. Jawaharlal Nehru, I fear, tries to find out our stand from the prejudices he has developed in respect of the various personalities who have publicly associated themselves with the Swatantra Party and he has not cared to study the principles accepted by the Party as constituting its stand. Our 'stand' is not to be guessed by psycho analysis but is to be seen in the principles we have agreed to and set out for public information with more clarity and precision and with greater brevity than one can find in any of the Congress pronouncements, or the Prime Minister's speeches. If this procedure were followed, it could be seen that the Swatantra

Party is not intending either to 'perpetuate' anything or leave the big industrialists 'to do what they like'."

Though Swatantra Party could not become an All-India Party which could capture a large number of seats, still it could capture enough seats to serve the purpose for which it was formed. It proved to be an effective opposition party.

Two main political ideologies—Democracy and Communism—govern the world. The majority of mankind lives either in a democratic State or under a totalitarian regime. In between these two types of present-day submission of people to authority, there is an amorphous mass of humanity whose fortunes and freedom waver at the whim and fancy of dictators, their lives a pawn in the power-mad antics of autocratically-minded despots who desecrate the most cherished dreams and aspirations of their people. These tyrants are fortunately few and far between, still they make their presence felt in the precarious existence of the people under them whom they seek to convert to their own concept of life and liberty. Rajaji took a significant part in the emancipation of India and leading it on the road to democracy. His deep study of free nations of the West and his experience of British rule in India made him welcome the advent of Swaraj to his own people. This resulted in Rajaji becoming a doughty defender of democracy throughout his life.

The existence of party system is an essential feature of a democratic State. Whether there are a few parties or many, whichever wins the vote of the majority of the electorate, becomes entitled to form the government. The great democracies of the West, the United States and the Great Britain, have only two parties and three parties respectively. The USA has the Democrats and the Republicans contending for power each time an election is held for the Presidency once in four years. In England there are three parties, namely the Conservative, the Liberals and the Labour Party; the real contestants for the assumption of authority to run the administration are only the Conservative and Labour Parties at the present time, for the Liberals are too few to have an effectual voice in the choice of forming government. In the French Republic, however, there are a large number of parties with the result that there is very often a coalition government. But one thing is certain that a party with a brute majority tends to forget the principles of democracy and to pursue policies that smack of pure totalitarianism, snapping its fingers at the rights and liberties of the individual citizen. Such a government embarks on programmes intent on perpetuating itself in power as long as possible.

"To make the personality of the ordinary men creative," observes Harold J. Laski, "it is necessary to build the conditions within which creativeness is possible. That can only happen when ordinary men are made to feel significant, and in the absence of liberty and equality, we cannot hope to achieve this. Where there is in a community the absence of those factors, which make the interests of men so differently considered, there is likely to be the means at hand of the development of personality. The enforcement of equality by the State has the great merit of promoting freedom by preventing the private person from the exercise of forces for his own ends. By force I do not mean necessarily physical violence, but the use of a differential advantage to hinder another from the opportunity to be the best he can."[9] Therefore, it follows that the individuality of man in a democratic State should not be overlooked, but given all opportunities to attain the fullest growth. The majority should not make use of its numerical strength to force its decisions down the throats of the minority. Rajaji speaks about the dangers of a massive majority with reference to the Congress Government and its ruthless march over the Fundamental Rights of the people against the letter and the spirit of the Constitution:

"This is not the first time that a massive majority has done what is wrong. Do not a massive majority of Jews insist on and obtain the Roman Governor's order for the crucifixion of Jesus Christ. A massive majority of MPs in the New Delhi has now resolved to destroy the Fundamental Rights, which our founding fathers had thought was necessary for the welfare and freedom of our people. As *The Hindu* in the course of its leading editorial of 6 August, 1971, said 'The sweeping powers conferred on Parliament by the 24th Amendment to tamper with all the rights, including the right to freedom of speech, of freedom of association and of religion, are such that it is not surprising that some leaders of minorities and even some committed socialists have been alarmed over the grim possibilities'."[10]

Not only had the Congress Party cornered power to a dangerous degree, callous to the welfare and interests of the common man, it had also been successful in the distribution of its favours only to such of those, who contributed handsomely to its party funds and enabled it to continue in power, without a moral right to do so. This it had done, by what Rajaji called the establishment of the "Permit-Licence-Raj".

Rajaji brands the insidious attempt on the part of the State to involve itself in an unholy competition with the individual citizen in various fields of his life and activities by the comprehensive term "Statism". That is, State replacing the citizen in healthy nation-building efforts and endeavours at the expense of his progress and prosperity, often crippling him with restrictions and cutting off the wings of action in the adventure of life and the achievement of real liberty. Rajaji considers that this statism is but the first step towards a totalitarian regime in the offing and borders dangerously on Communism. The leaders of the Congress Party equated statism with socialism, says Rajaji, but he reminded them that even Jawaharlal Nehru, with all his enthusiasm for socialism, drew a distinction between Socialism and Communism and denied that the ideology of the one is the same as that of the other. "I am deeply convinced," declared Jawaharlal Nehru, "that the methods in certain communistic societies, that is, too much coercion and suffering, are not the right methods."[11] Long before he became Prime Minister of India, he had visited Soviet Russia". Though on his first visit to Russia, he had found much to admire in Communism and the communist way of life, he was greatly disillusioned soon afterwards. He discovered that millions had no rights worth the name of freedom and had been made mere pawns in political philosophy that wrought havoc with the dignity of man and his individuality. "I had long been drawn to Socialism and Communism," he said in his Autobiography published in 1936, "And Russia had appealed to me. Much in Soviet Russia I dislike—the ruthless suppression of all contrary opinion, the wholesale regimentation, the unnecessary violence (as I thought) in carrying in and carrying out various policies."[12]

Rajaji's attack on Communism is based, as much on his boundless belief in a beneficent God, as the totalitarian ideology is a denial of a Divine Being, the Creator of the Universe. Communism desires to wreck the soul of man on the rock of regimentation, running him without any hope of redemption. Communism seeds to destroy the individuality of man and direct his incomparable mind into the dark caverns of despair and frustration of spirit. Communism wields the ancient trident of tyrants—fear, frustration and fetters—in order to bring millions into submission under its heel and crush all that is noble in a human being into nothingness and a negation of the divine destiny of man. Lastly, Communism confines the free wings of man yearning for light and air, song and sunshine, within the iron bars of a cage constructed out of a callous indifference to the infinite capacity of human beings in their quest for Heaven!

Rajaji's love for democracy and his loyal adherence to its principles is refreshing in lurid contrast to the retrograde attitude of then ruling party since Independence. While offering lip service to democratic ideals, they did all that they could to discredit democracy and deny its benefit to the common men. He deprecated strongly the prevalence of one party rule for a continual period of nearly quarter of a century, leading to an unhealthy trend towards totalitarianism in the administration. "The successful working of parliamentary democracy", said Rajaji, "depends on two factors: first, on a board measure of agreement among all classes of citizens about the objectives of government; secondly, on the existence of a two-party system, in which each of the big political groups possesses effective and continuous leadership and is strong enough to take over the responsibility of government when the majority of the country's voters wish it. If political opinion does not succeed in crystallising into two fairly even balanced groups, the semblance of democracy may survive but real parliamentary democracy will not be there. When one party always remains in power and dissent is dissipated among unorganised individuals and relatively insignificant groups, which cannot coalesce, government will inevitably become totalitarian."[13] Rajaji had correctly diagnosed the disease in the body politic that was undermining the rights enshrined in the Constitution. The Congress Government had been treating the basic law of the land as if it were a plaything or a lifeless block of wood for them to hew and hack at in order to shape it for their political purposes. Rajaji laid emphasis on the need for an effective Opposition, which should fulfil its *dharma* of criticising the rulers when they went wrong. If the opposition should be weak or fail to do its duty by the people, the country will be ruined, not only materially but, also in terms of moral fall from the high ideals, whose adoption by Gandhiji exalted it in the esteem of world nations, bringing freedom in its wake.

"A strong opposition is essential for the health of democratic government. In a democracy based on universal suffrage, government of the majority without an effective opposition is like driving a donkey on whose back you put the whole load in one bundle. The two-party system steadies movement by putting a fairly equal load into each pannier. In the human body, two eyes and two ears enable a person to place the objects seen and heard. A single-party democracy soon loses its sense of proportion. It sees but cannot place things in perspective or apprehend all sides of question. That is the position today."[14] Rajaji was right in his opinion that the various political parties in India were frittering away their energies in opposition to the Congress Party, without combining together to pull it down from its position of supremacy. "It is good to have a giant's strength but tyrannous to use it so," said Shakespeare. Rajaji echoed the wisdom of the Bard of Avon but the Congress giant was going on merrily without caring a jot or little for the opposition parties, stabbing democracy in the back in the name of socialism and the building up of a socialistic pattern of society. "The dogs bark but the caravan passes on," said Sir Samuel Hoare, the Secretary of State for India, in the British Parliament, at the height of the freedom struggle, deaf to the demands of the Indian people for *Swaraj*. The Congress rulers of India had become worthy descendants, gloating in their hearts at the discomfiture of the opposition parties!

Rajaji was not against Hindi. In fact he had been a propagandist of Hindi during the freedom movement. But he was against the haste with which Hindi was being imposed. He was all for retaining English. He was not in favour of Hindi to replace English. He wrote in the *Swarajya* that, "Unless we wish to hasten disintegration and become the laughing stock of the world, we must put the Hindi claim to replace English in cold storage and meanwhile let the *status quo* be the rule. We can carry on as we carried on from 1947 to 1950 with the dignity and world respect which we then commanded."[15]

Regarding the need for retaining English indefinitely he said, "I am one of the oldest now among those living that were in the fight for political emancipation. I have to be a rebel in the matter of language as I have been in the battle to recognize the natural laws of production of wealth and to give up the vain effort to ignore them. I have now to appeal to the University professors and the youth learning under them to revolt against what the politicians are seeking to do to uproot the language of study, reference and instruction in modern enlightenment, viz., English which has rooted itself in the universities and to rely on a new mixed plantation which it is hoped will efficiently take its place. This is a war against reality, doomed to failure and disasters of great size. We cannot write-off the history of two hundred years at the call of nostalgia or love of classic India. It was not merely a period of foreign rule. It was a period of great world changes, a period of new enlightenment, and technological progress alongside of foreign rule.

Rajaji condemned the manner in which the Congress Government treated the Indian Princes regarding the payment of their privy purses. He was terribly against the deprivation of the privy purses of the princes by amending the Constitution. He said: "To repudiate one's own obligations after getting full performance from the other side, is disgraceful and unwise". He added "The immorality of the attempt to deceive the rulers who became Indian citizens and voters in the Union of India, on the basis of the assurances given and after the fall performance by the rulers of their part of the contract, is something that cannot be described except by the strongest words of condemnation."

In 1962 Rajaji led a three-member Mission for World Peace to the United States. The other two members were R.R. Diwakar and B. Shiva Rao.

In a letter to the *New York Times* Rajaji spelled out the object of his Mission. "I am in the United States as a member of the Gandhi Peace Foundation Delegation that arrived in Washington on September 27, 1962 on a mission to try to secure the immediate cessation of nuclear tests. The present deadlock appears to be the result of the Soviet Government not accepting the offer of the U.S. Government to stop all tests in the atmosphere and the sea without any conditions, if the Soviet Government does the same. The Soviet Government has insisted that any agreement should cover also underground tests, over which there is an unresolved difficulty about verification and inspection. It will not consent to make a beginning with an agreed cessation of the tests over which there are no difficulties, as to verification and inspection. The position, therefore, is a needless stalemate in respect of even what could be immediately done to the great advantage of the health of humanity. I feel, under these circumstances, that the one way out of the stalemate is for the United States to take unilateral action, putting its offer into practice and promoting a prohibitory resolution in the UN for the cessation of all atmospheric and undersea tests and calling for a speedy and successful termination of the negotiations ever that cessation of underground tests .... If it commands massive, support in the UN, it is not likely that any member, however powerful, would violate its terms. In any event, I believe that the time has arrived when the moral sense of the peoples of the world should assume a dynamic form and compel the cessation of the continuous contamination of mankind's environment."[16]

Rajaji made a deep impression on John F. Kennedy, the President of USA, B. Shiva Rao, a member of the Mission delegation, later recorded.

"In 1962, it was my privilege to accompany Rajaji to New York, Washington and London on a great mission, the suspension of nuclear tests as a preliminary to disarmament. I remember vividly, as though it was yesterday, one episode our meeting with President Kennedy. We had been warned that it was very busy day for the President, and he could spare only 20 minutes. But Rajaji's presentation of the case was so convincing that Mr. Kennedy sat fascinated. The minutes sped well beyond an hour, but the President ignored his other engagements to listen to a deeply moving appeal by Rajaji on behalf of defenceless humanity.

"Finally, as the interview terminated, Mr. Kennedy gave him the assurance that something would be done : and later he told one of his officials present at the interview that never before in his life had he felt the impact of ancient wisdom as he had done from Rajaji."

Before leading the Peace Mission Rajaji had delivered numerous speeches and written many articles advising the big powers to stop nuclear tests in the name of humanity. The Delhi Peace Council had brought out a book in 1957 entitled *Mankind Protests* containing a collection of Rajaji's speeches and writings on the subject of nuclear warfare. A Russian translation of this book appeared in Moscow in 1958. In 1960 National Book Trust had brought out a book *The Voice of the Uninvolved* containing Rajaji's speeches and statements on Atomic Warfare and Test explosions. Rajaji felt that the great powers that had secured the secret of nuclear power and developed deadly weapons had no right to condemn unknown, unnamed and unborn children to mutilation, torture and death merely by experimenting with weapons for its own defence. This may even lead to the annihilation. He wrote in the *Preface to Mankind Protests*, "It is idle to expect the world to give up its opposition to present poisoning because the tests may lead to the evolution of more manageable nuclear weapons. Whether it is one big blast or one hundred smaller blasts, the nuclear programme is a delusion and a snare for everybody. Once the game starts, the frustrated party has no escape from using its power to strike effectively, and that means the last, or last but one step to annihilation."

Rajaji continued to guide his countrymen and the world at large through the columns of *Swarajya*. Ved Mehta gives a pen picture of Rajaji. Ved Mehta had gone to Madras to see Rajaji. Rajaji, it may be mentioned, was eighty-eight years old at that time. Rajaji told Ved Mehta, "I am not so good in my ears and my eyes are also bad. I had cataract and I still have severe myopia. My sense of smell is good and my sense of touch is very good. But my digestion is very bad. I am an old man, and sometimes my health is not good, but I am a happy man. According to my own lights my mind is good, but the Congress Government in power in Delhi does not seem to think I am very clear headed. All my life I have been engrossed in some kind of work. I have never required the usual tonic of recreation. Most of my work now is for *Swarajya.* Every week there is the front page article to write and my personal notes to the reader, 'the Dear Reader' column; and sometimes there are other opinions and observations I want to write down for *Swarajya* readers, and so I contribute a third article. *Swarajya* has a circulation between fifteen and eighteen thousand, but it deserves to have a circulation of a hundred and fifty thousand."

Rajaji fell ill in November 1972. He breathed his last on 25th December 1972. A few minutes before his death he asked his relatives who appeared to be sad and depressed, "Why are you doleful? I have not made an appointment with Yama yet." When he asked him how he was, Rajaji replied, "I am happy." I am happy were his last words. He, of course, was happy because he had played his part well but the humanity was sad over his demise.

REFERENCES

1. Kaamadhenu is something that gives ones whatever he wants.
2. *Swarajya*, 21 October 1961.
3. *Illustrated Weekly*, 14 August, 1960.
4. *Swarajya*, 17 October 1963.
5. *Ibid.*, 14 August 1971.
6. *Ibid.*, 21 October 1961.
7. *Ibid.*, 10 February 1968.
8. *Ibid.*, 25 July 1959.
9. Laski, Harold J., *A Grammar of Politics*, p. 170.
10. *Swarajya*, 14 August 1971.
11. Mende, Tibor, *Conversations with Mr. Nehru, op. cit.*, p. 32.
12. Nehru, Jawaharlal, *An Autobiography* (London: The Bodley Head), 1958, p. 361.
13. Felton, Monica, *I Meet Rajaji* (London: Macmillan), 1962, p. 52.
14. *Ibid.*
15. *Swarajya*, 1 April 1967.
16. *New York Times*, September 1962.

5

NATIONALISM TO INTERNATIONALISM

"We stand at the end of an era and on the threshold of a new period of history. Standing on this watershed, which divides two epochs of human history and endeavour, we can look back on our long past and look forward to the future that is taking shape before our eyes. Asia, after a long period of quiescence, has suddenly become important again in world affairs. If we view the millennia of history, this contact of Asia has played a mighty role in the evolution of humanity. It was here that civilization began and man started on his unending adventure of Life. Here the mind of man searched for Truth unceasingly and the spirit of man shone out like a beacon which lighted up the whole world."[1]

A deep and abiding interest in international affairs marked Rajaji's public life and political career. Being one who had taken a substantial part in India's struggle for Swaraj and having also been responsible in raising her to a position of prestige and dignity among world nations, he felt after retiring from the office of the Governor-General and the other distinguished positions in the Central Cabinet at New Delhi, that he should not devote himself only to home affairs but should also engage himself in problems that agitated other countries. His keen intellect and uncanny gift for probing into the heart of a problem, however, complicated and intricate, enabled him to take a balanced and practical view while others, who were swayed by passions and partnership, found themselves entangled in a mesh of Machiavellian intrigue and counter intrigue. Rajaji had accepted Mahatma Gandhi's magnificent ideal of the spiritualization of politics. He, therefore, suggested to the statesmen of the other countries the application of the same ethical standards and moral principles when they were confronted with difficult and delicate situations. His words may have fallen on deaf ears and his advice may have been unheeded but such apathy to the conscientious advice of a thinker of rare acumen and remarkable integrity has landed the world in the unhappy situation in which it finds itself at the present day. He always considered that "the rarer action is in virtue than in vengeance", especially where the interests and welfare of large masses of people in conflict were considered. Being a great humanitarian first and last, Rajaji always thought of the common man, whether it was in India or in Indonesia, the Great Britain or Germany, the United States of America or the Union of Soviet Socialist Republics. He always prayed and worked for the ushering in of a day when there would be a World Government in which all nations, big and small, participated to build a paradise on earth for the poor and the rich alike!

Before we consider Rajaji's views on international problems, it is worthwhile to compare his opinions with those of contemporary leaders in India who guided her destiny. The most notable amongst them was Jawaharlal Nehru for whom Rajaji had a great regard and affection but from whom he differed strongly on his policies and programme of action. His chivalrous attitude to India's first Prime Minister was such that when he passed away on 24 May 1964, Rajaji cried out in agony: "He was the most civilized amongst us all. The country can at this juncture afford to lose me a thousand times but not Jawaharlal Nehru!" A tower of moral strength and support to Jawaharlal Nehru when Rajaji was Governor-General of India and later when he was Home Minister of the Central Government till 1952, he found himself unable to accept the methods of administration and the motives behind what Jawaharlal Nehru did at home and abroad. The most important subject, perhaps, on which he differed from Jawaharlal Nehru was the principle of non-alignment, which he propounded and practiced when dealing with foreign countries. Jawaharlal Nehru summed up non-alignment in these words:

"Non-alignment, despite the negative form of the term, is a positive concept and we do not propose to have a military alliance with any country, come what may. The moment we give up the idea of non-alignment, we lose every anchor we hold on to and we simply drift."[2]

This principle of non-alignment was foreshadowed in the very first speech he delivered as the Head of the Interim National Government formed in September 1946. Broadcasting the foreign policy of the Government of India he said:

"We propose, as far as possible, to keep away from the power politics of groups, aligned against one another, which have led in the past to world wars, and which may again lead to disasters even on a vaster scale. The word, in spite of its rivalries and hatreds and inner conflicts, moves inevitably towards closer co-operation and the building up of a world commonwealth. It is for this One World that India will work, a world in which there is the free co-operation of all peoples, and no class of group exploits another."[3]

That pronouncement as India's Foreign Minister made Jawaharlal Nehru justly famous as the exponent of the policy of non-alignment not aligning with either of the two power blocs of the West. The reaction of some nations to this idea was retrograde while others applauded non-alignment, as the only method of avoiding a conflagration that would destroy all that was precious in civilization. Jawaharlal Nehru considered that this policy ensured the continuance of human freedom, the dignity of the individual, the happiness of mankind in general and last but not least, the harmony of good relations between big nations and small nations.

The basis of the policy of non-alignment was Jawaharlal Nehru's staunch belief in non-violence, which was one of the twin weapons wielded by Mahatma Gandhi in his fight for India's freedom. He had become the favourite lieutenant of Gandhiji in his campaign for India's liberty and had accepted the ideals of truth and non-violence as essential for success. Mahatma Gandhi had declared on a memorable occasion that *ahimsa* and truth were as his two lungs and that he could not exist without either of them. Jawaharlal Nehru's acceptance of the principle of *ahimsa* or non-violence did not mean that he would advise people to run away when attacked by a ruffian. Nor did it make him a pacifist. 'I am not a pacifist', he declared in 1956. "I do recognize that under certain circumstances, one has to fight. It depends less on theory than on the background of the people on what they can do. Even Mr. Gandhi, who was a great pacifist, always said that it was better to fight than to be afraid. He meant that you must not surrender to evil, to basic evil, and that you must preferably fight in a peaceful way. If you cannot do that, then fight in the military way. But don't surrender to evil."[4]

Jawaharlal Nehru worked for world peace without being pacifist. His aim was the implementation of Mahatma Gandhi's ideals by pursuing the path of non-alignment. Each of the big powers were intent on dominating the world with their own ideology without bothering about what might happen to mankind in case of an armed conflict in which they would use their most lethal weapons to destory their enemy. They forgot that success or defeat for either party meant the destruction of the glorious heritage of humanity, its precious culture and incomparable civilization, leading to the ruin of millions of innocent people, who pine for peace and happiness.

Rajaji was an avowed pessimist regarding the success of the policy of non-alignment. He felt that India should take a definite and determined stand on the side of the great democracies of the West like the United States, Great Britain and France in their opposition to the totalitarian doctrines and regimes that were casting their sinister shadows across the world. The freedom that India had won after a long and arduous struggle with Britain would be mere mirage if the minds of men were infected with the negative policy of non-alignment, instead of their lending active support to the ideals of democracy. Rajaji wanted to stem the creeping tide of Communism and save the simple and credulous masses of his country from its vicious grip. The idea of non-alignment does not mean that India should be anxious to purchase "peace and security at any price," which would surely destroy her independence. Hence, Rajaji declared that there was a positive need for the revision of India's foreign policy in order to ensure the fulfilment of her great ideals. The goal of non-alignment, however laudable in terms of achieving world peace, should be deemed to have no meaning—a mere delusion hugged fondly without any redeeming feature and after

China had invaded India's northern border while Soviet Russia merely looked on us as an indifferent spectator without taking any concrete steps for preventing China's aggression on India. This clearly proved, said Rajaji, that the doctrine of non-alignment would be an excellent policy for India if her independence could cover the economic needs as well as the defence requirements of the country arising from a national policy of non-alignment. "So far as India is concerned, after the Chinese aggression and after the disappointment in our expectation that Russia would interfere, it is clear that our attachment to non-alignment is reduced to a meaningless superstition. We tried hard to get the Soviet Union to see our plight and appealed to her to save our non-alignment, but Russia was not willing or was not able to induce China to call off her aggression. To believe that Russia will, later on help us to retain our frontier...is really to fool ourselves."[5] The utter futility of his fallacious policy of non-alignment, said Rajaji, was evident from the fact that the then ruling party was more intent on defending non-alignment than on defending the country.

Rajaji was not a mere carping critic of the Congress Government or its policies as put forward by Jawaharlal Nehru and implemented in its relations with foreign powers. He would not only find fault, when necessary, with its domestic or foreign policy but also offer constructive suggestions to wean them from their wrong path in the true interests of the welfare of India and her people. The policy of a neutral strategy that is the basis of non-alignment exposed its weakness and utter lack of substance when China attacked India in the Himalayan region. In support of this argument Rajaji quoted with approval an article by Professor Werner Leir in the London monthly, *Eastern World*. The learned writer said: "India came within the sphere of a powerful nation's imperialistic drive. One of the conditions for successful neutralism was the absence of such an entanglement. India ignored this difference even after the grabbing of Tibet, the construction of a road through the northern tip of Ladakh in 1956, the building of a strategic roads and airfields along the whole Himalayan border people by subversion and propaganda. Yet India steadfastly adhered to neutralism and rationalized her attachment to it. Had India more modestly treated neutralism as a strategy and limited it to where it was useful, and not adhered to it as a doctrine or a permanent way of life, she might have anticipated China's aggressive intentions and prepared for them. China, always cautious, might not have attacked India even as she has not attacked so far any other nation on her borders which is protected by an alliance."[6]

The only way of India encountering any further aggression by China or any other hostile Power was entering into an alliance with another major world power, rather than embarking on an arms race. It might be difficult to have a profitable alliance, Rajaji admitted, due to India's past indifference to offers of friendship. But an attempt at obtaining aid was sure of success as there were basic bonds of culture and history between India and the Western democracies. "The way to peace and progress," argued Rajaji, "is alliance and its potential power, rather than actual hardware. An armament-buying race to replace the Himalayan wall is not the way to progress or peace or strength, but a shortcut to bankruptcy and collapse. Alliance has assumed for us greater significance now than before. The potential power of friendship secured by diplomacy is far safer and richer plan of defence than a heavy arms build-up. It calls, however, for courage and humility and the discarding of old rigidities and semantic-based fallacies—a difficult demand to make of the present regime in Delhi."[7]

Internationalism in politics leads to the concept of a world government and the quest for world peace—two ideals for which Rajaji always engaged his efforts and energies. But there is a higher and nobler concept of internationalism in the realms of civilization and culture, literature and art. When one remembers that Dr. Alfred Nobel, the Swedish scientist, has built the greatest monument to his memory by founding the Nobel Prize of Science and Medicine, Literature and Peace, etc., open to all eminent men at the top in these fields of achievement without distinction of caste or creed, clime or country and that the Prize has been awarded year after year to the greatest men, it is no wonder that an intellectual of the front rank among the world's intellectuals, should shed his insularity and plead for the adoption of an international language like English as the *lingua franca* of Independent India. "I want the winds from every side to be blown across my country," said Mahatma Gandhi who fought against insularity of every kind being injected into the minds of men by the so-called leaders in the name of patriotism. Gandhiji loved India and her people intensely but his love was not parochial, extending beyond its borders to all corners of the world and embracing every human being from China to Peru, even in the days before he led his country to liberty. A true follower of his Master, Rajaji was an internationalist in the matter of language also and wanted his countrymen to adopt English as their own language while cherishing their own languages and their literature, ancient and modern. He fought for the retention of English against the diehard tribes of politicians who plumped for Hindi as the official language for the purpose of administration. English had brought unity to the various parts of the subcontinent, he said, and it was because of our knowledge of English and its vast literature and thought, that Indians were able to compel the British to withdraw from India voluntarily. Why then, he questioned, should we banish English from our midst, which has brought us so many blessings in its wake? He boldly declared that the giving up of English would bring disaster to the country, it would only balkanize India and break it up and its people into various small groups of selfish men, with parochialism ruling their minds and not patriotism of the noblest kind of moving their souls to yet higher endeavours and efforts. "Unless we wish to hasten disintegration and become the laughing stock of the world, we must put the Hindi claim to replace English in cold storage and meanwhile let the *status quo* be the rule. We can carry on as we carried on from 1947 to 1950 with the dignity and world respect which we then commanded."[8]

English has been responsible for peace and progress in India and it was built itself into the hearts of millions on a solid foundation of rock, which cannot now be shaken except with disastrous consequences to the country. How can we forget the fact that the treasures of English literature have moulded the minds and stirred the souls of the noblest and greatest sons and daughters of India during the last hundred years? It is undesirable as well as impracticable to think of giving up English at any future date in favour of Hindi, which should only be given the status of a regional language. There is absolutely no room for compromise with regard to the necessity of retaining English indefinitely. "I am one of the oldest now among those living who were in the fight for political emancipation. I have to be a rebel in the matter of language as I have been in the battle to recognize the natural laws of production of wealth and to give up the vain effort to ignore them. I have now to appeal to the University professors and the youth learning under them to revolt against what the politicians are seeking to do—to uproot the language of study, reference and instruction in modern enlightenment, viz. English which has rooted itself in the universities and to rely on a new mixed plantation, which it is hoped will efficiently take its place. This is a war against reality, doomed to failure and disasters of great size. We cannot write-off the history of two hundred years at the call of nostalgia or love of classic India. It was not merely a period of foreign rule. It was a period of great world changes, a period of new enlightenment, and technological progress alongside of foreign rule."[9]

It is not Hindi or any other regional language that gives dignity to the nation continues Rajaji, but India's classical language, Sanskrit, in the golden past. But at the present day he asserts that India's classical language is English, which binds all Indians with a strong rope. "It is a medium which is a sustaining blood-vessel of nutrition and national unity." Rajaji considers that unity is not achieved by regimentation either in language or in dress or in thought. He believes that variety is the spice of life and the various languages in the country should all develop in their own way, while fostering the feelings of friendship with those who speak other languages and fusing the people ultimately into one united nation in spite of their diversities of thought. His opponents thought that he was against Hindi, forgetting that he was one of the most ardent propagandists for Hindi in the South of India, nearly half a century ago. Even at present day he is not against the making Hindi restricted to its place as a regional language but he stoutly denies Hindi the supreme place in India's life and thought at the expense of English. "I still believe in making Hindi compulsory", he frankly confesses, "as a subject in schools, although I am against its replacing English as the official language of the Union of India. People think that this is a contradictory attitude. But it is not. You don't bring unity to the country by imposing uniformity. Instead you make the people split by creating bad feelings between them. Distribution of wealth is good if you believe that wealth itself is good. But language is a more important possession than wealth."[10]

More than six decades have passed since the defeat of the Axis Powers in May 1945, when the Allies of the democratic nations and Soviet Russia struck the death-knell of Hitlerism and Fascism. But their victory has not all been a blessing to the world. It has left more baneful results than beneficent effects on mankind. For, with the proud consciousness of having humbled their enemy, the Big Powers of the West have begun to pile up atomic and nuclear arms, each suspicious of the others, each lusting for supremacy, each absolutely callous to its moral duty to see that its unbridled ambition does not lead to the tragic destruction of humanity and its priceless heritage of ages. A cold war, more threatening to the world's civilization and culture, has developed between the democracies of the West and their totalitarian counterparts. This cold war not only affects the peace and happiness of millions in their own countries but also involves in its wicked grip the innocent and simple masses in the underdeveloped countries of Asia and Africa. India, belonging to the latter group, has to suffer the consequences of being within the sphere of influence of the insidious propaganda of the totalitarian Powers and their programme to achieve supremacy for themselves and their doctrines. Rajaji was one of the few statesmen of international reputation who had boldly condemned this attitude of authoritarianism in spite of their professed desire for establishing peace and concord among all nations. In order to dissolve the foul atmosphere of deadly doubt and dangerous possibilities to the security of other nations, Rajaji took the initiative in addressing a letter to the General Secretary, Khrushchev, in which he pleaded strongly for the unilateral abandonment of nuclear weapons by the Soviet Government: "It is needless for you to point out to me the lapses of the Western Powers. I know them all and I have been pointing them out publicly," Rajaji continued, "but a supreme moment has now arrived when your Republic can attain undying glory, a great and historic step whose moral force will be irresistible. The pattern of competition in the future would change from one of destruction to the unravelling of the mysteries of nature and the promotion of human welfare."[11]

The response of Mr. Khruschev to Rajaji's letter contained a reference to the discussion he had with Rajaji on the preservation of world peace even during a Cultural Festival arranged especially for the Russian Premier. Khrushchev expressed his views on nuclear disarmament and confessed that he had his own doubts about the feasibility of unilateral action, and its moral "chain reaction" on other Great Powers, if Russia should take the initiative in the matter by renouncing the use of nuclear weapons. Khrushchev concluded by stating that after a great and prolonged discussion on the subject, the Soviet Government was reluctant to take upon itself the unilateral obligation. Rajaji sent a rejoinder soon after to Khrushchev appreciating the latter's anxiety to end the cold war. "The very pith and substance of the cold war is suspicion," said Rajaji in his letter to the Russian Premier. "It can be put an end to only by one party or the other beginning with one's own unilateral step...this necessarily involves risk. But any process that does not involve risk cannot end the cold war. The cold war is going on, developing a terrible risk by itself. We have to compare one risk with another. Ending the cold war means suspending our suspicion itself. We are therefore in a terribly vicious circle. The more I think of it, the clearer it is to me that we cannot hope to end this cold was before it bursts into flames."[12] An equally long rejoinder to this communication of Rajaji from Khrushchev mentioned that in the hope that other nations would emulate their good example, Russia had largely reduced its armed forces. But, continued Mr. Krushchev, they had been disappointed. "We are convinced," concluded the Soviet statesman, "that if the unilateral action undertaken by us now were supported by similar efforts of the Western Powers, it would lead not only to the improvement of the international situation but would open the way to newer steps with the purpose of deliverance of humanity from the armament race and from the menace of atomic war."[13]

Nearer home the one absorbing topic of immediate interest to Rajaji in the international field was the relations of India and Pakistan *inter se*. Though he believed that in the interests of both countries, their separate existence as independent States was inevitable, he could never reconcile himself to the idea that they were two different nations, who had parted each with a separate status of sovereignty. On the other hand, he considered that they were alike two brothers who had only partitioned their ancestral estate, inheriting their common legacy of culture and civilization built up by them down the centuries of their co-existence in a common country. Consequently Rajaji argued that they should never look upon each other as enemies or rivals in any field but as partners intent on working for the welfare of the masses both in India and Pakistan.

Rajaji reiterated need for permanent goodwill between the citizens of the two independent States. Even to the extent of being misunderstood, he pleaded again and again that the relations between erstwhile brethren should be marked by a spirit of "forgive and forget", whatever were the tragic events that had happened after the partition of India. "It is an intolerable evil," he wrote, "that two masses of human beings should have feelings of kindly brotherhood towards each other but that they should be prevented from acting according to their nature by the governments that they respectively have appointment for themselves, and worst for all, that the two people should feel wholly impotent about it. Both the government and the people submit to the paradox as inevitable. The consequence is that the governments make a deliberate effort to remove the paradox by bringing people to think as their governments do! The governments have enormous power and effective instruments for this conversion. The cure in this case is certainly worse than the disease, for it substitutes friendliness by suspicion, fear and hatred."[14]

The tragedy that followed on the heels of the Tashkent Peace Agreement between India and Pakistan after their war in 1965, namely, the death of Lal Bahadur Shastri, the Indian Prime Minister, saddened Rajaji immeasurably. He compared the passing away of Lal Bahadur Shastri soon after the victory for peace at Tashkent with the loss of General Wolfe after winning Quebec for Great Britain in 1759. Rajaji condemned the attitude of some elements in the two countries that could only think in terms of conflict and victory in war. "A seed has been sown at Tashkent," said Rajaji, "which is small like the tiny seed of the Nyagrodha tree which the *Rishis* in the *Chandogya Upanishad* asked Swetaketu to observe. Yet in this tiny seed is contained the great and precious wide branching tree of friendship and mutual co-operation between two big nations—a Nyagrodha tree of co-operation, which will make South Asia great and powerful and become a strong pillar of civilization that will attract the wonder and love of the nations of the world."[15] As a consequence of the Tashkent Agreement, Rajaji foresaw an era of neighbourly goodwill and mutual benefit between India and Pakistan in a common economic effort, founded on secularism and toleration of each other's faith, and without any inconsistency with their own individuality or sovereignty or affecting their national pride.

Not being a narrow-minded nationalist but a great believer in the good wrought by working for international amity and goodwill, Rajaji was equally emphatic on the sheer necessity for observing the principle of co-existence. The world, as it is today, and as it has always been, contains many nations, many cultures, many civilizations and also ideologies of different kinds. The survival of man demands that there should be a sort of unexpressed but clearly understood basis of acceptance of different forms of thought so that men may continue to pursue their objectives for the common welfare of humanity. Man's progress in the scale of civilization down the ages has been indicated by the gulf that divides man from the wild nomadic life and those in a cultured and civilised society. Similarly, when the different nations of the world have to live together and bring their own contribution to enrich the heritage of humanity, they cannot afford to alienate themselves either in thought or in action from the others. This concept of common ideals and common endeavours in the interests of mankind led to the founding of the United Nations Organization in 1945 soon after the termination of the Second World War. Rajaji had firm faith in the efficacy of the principle of the co-existence of nations. "The human family which issued out of the same parents, whether according to religious traditions or according to the theory of the evolution of species, has developed different features and different complexions of the skin. So also it has developed different ways of life according to the differing physical and historical causes that shaped them, either through long years of slow change or through sudden revolutions. If we love and desire peace, we should learn to recognize the value of our neighbour's way of life (to him) as we value our own and tolerate one another without seeking to impose our convictions and manners on others. Co-existence is the name given to this tolerance and this wisdom, for this is wisdom, intolerance being folly."[16]

A man of many-splendoured vision and nature wisdom, Rajaji was also a great optimist, and had a boundless faith in the future of man. He therefore dreamt of a day when there would be a world government. At the present time the Big Powers are in the vicious grip of the cold war, endangering not only their own security but also the very existence of other nations whom they will not allow to develop in their own way, according to their traditions and culture. The statesmen of Western nations, who are responsible for this world insecurity by their callous race in armaments, must give up their folly and pave the way for the people of different countries, cultures and civilizations, to become one human family, intent on giving of their best for the benefit of mankind and having the noble ideal of working for the harmony and happiness of all the nations of the world. It may not be possible to establish world government so easily, but world peace is within the reach of man if only he makes an honest effort, considering no loss of power or prestige too great for that purpose, whether such giving up effects the big nations or the small, in the East or in the West. "The secret of world peace," said Rajaji, "is in doing away with all international hurdles and protective and revenue tariffs and making the commerce in commodities and of thought between nations completely free, as it was between the Chinese Empire and India in the old days. It may be a dream, but a dream that should be dreamed persistently. It would be easy to get all nations to exempt medicines and baby foods and the like from all custom duties. It would be then possible to bring other commodities gradually into the scheme of free commerce, until the human family discovers itself as one industrious people working in convenient centres, and competing with one another to improve quality and reduce costs of commodities without erecting walls against one another creating artificial scarcity and adding to the prices for the sake of government revenue. The budgets

of the nations and their tax systems will then suffer a sea-change which may not be fully visualized now but which will make for greater human happiness all around."[17]

REFERENCES

1. From Jawaharlal Nehru's Inaugural Address at the Asia Relations Conference held at New Delhi on 23 March 1947.
2. Nehru, Jawaharlal, *India's Foreign Policy, op. cit.*, p. 326.
3. *Ibid.*, p. 2.
4. Mende, Tibor, *Conversations with Mr. Nehru, op. cit.*, p. 79.
5. *The Hindu*, 23 December 1960.
6. *Swarajya*, 16 March 1963.
7. *Ibid.*, 8 June 1963.
8. *Ibid.*, 1 April 1967.
9. *Ibid.*, 19 August 1967.
10. Felton, Monica, *I Meet Rajaji, op. cit.*, p. 87.
11. *Ibid.*, p. 105.
12. *Ibid.*, pp. 106-107.
13. *Ibid.*
14. *Swarajya*, 27 August 1960.
15. "*The Onlooker Annual*", 1966.
16. *Swarajya*, 3 June 1961.
17. *Ibid.*, 17 December 1966.

6

INDIAN CULTURE

Rajaji was a deeply religious man with an abiding faith in God. Though he appeared to be engrossed in worldly affairs and spent all his time and energy in the rough and tumble of political life, he had anchored his soul in the Supreme. Like his Master all his activities were motivated by a ceaseless quest of the spiritual. Just as it was an accident that Mahatma Gandhi had strayed into politics, Rajaji had engaged himself from his younger days in the fate and future, the fortune and freedom of his fellowmen. Like Mahatma Gandhi he was in the world and yet out of it, with his eyes ever set for a glimpse of the Beyond. He was the ideal *grihastha*, who never ignored the responsibilities of family life. He fulfilled his duties towards his wife and children but was always in quest of the something beyond life and beyond death, which has baffled the wisest of mankind and has been the end and aim of their existence in their endeavour to merge themselves finally with the Eternal Spirit!

Men following all faiths from time immemorial accept that religion is the pathway to God. There are, no doubt, millions of men who do not believe in the Almighty and who have no faith in the efficacy of religion to lead them towards a land of bliss where the Omnipotent Creator rules with love over all His creatures. One can have no quarrel with such people for they have missed the path and are struggling alone in darkness and despair, unable to see the Light of God that is manifest everywhere and in everything, whether it is a microbe or a monster, a whale or a vermin, a mountain or man. "Religion is essentially faith in the reality of God's rule," says Rajaji.

"Through this faith man acquires limitless courage and capacity for suffering and sacrifice. They acquire a sense of equality and the oneness of the human family, and, indeed, the oneness of all life."[1]

Just as men have to depend for food for their existence from day-to-day, spiritual food in the shape of religion is equally essential for their survival if they are to be differentiated from the mere animals in creation. What distinguishes man from animals is his consciousness that he is a creature far advanced in the scale of evolution from the living things whose be-all and end-all seems to be a meaningless continuance of their species. Men realise that there is supreme purpose in their birth and in their death. "The Universe of matter with its unchangeable laws does not explain itself. That it exists does not satisfy the human mind, although the atheist may refuse to give further thought to it. At that point beyond which reason cannot operate, Faith begins. This is Religion. It covers the ground which science cannot explore. In particular, it gives rise to commandments as to how men should act towards one another and among themselves."[2]

He looks upon atheists as men who refuse to look beyond their noses, as it were. They pride themselves on their defiance of God and God's rule. He looks with horror at the godlessness that prevails among people all the world over today. It is this godlessness that is the parent of the suicidal policy pursued by the Big Powers in their desire to dominate the world, though in the process the whole of mankind may be completely destroyed. He considers this godlessness worse than atheism. Our intellectuals, children of modern knowledge, have not only developed into mere atheists but also have also become victims of godlessness with all its grim consequences. "It is this godlessness difference which we must cure for all our maladies really result from it. Atheism is passive transfer the word non-cooperation with their idea of God. Godlessness is not only that but a revolt against the moral rules of conduct that have issued out of all religions."[3]

Rajaji desires people to realise that men and women come into being because of God, they exist because of God, and God ordains their end. He wants them to remember that it is God who moves the working of our minds. It is the mind of man that has enabled him to have the rare power of survival among all created things, supreme and unsurpassed, their lord among all creatures having life. This is not a mere physical capacity to conquer all obstacles and come out triumphant overcoming all trials and tribulations. This incomparable power of man, declares Rajaji, is a spiritual power and a proud privilege not possessed by any other living thing. It may be asked from where this power comes to man, how he happens to be endowed with it, who gave such outstanding potency to man? Rajaji answers in one simple word, as all the wise men of the world answered: "God!" We may not be aware of His existence, we are conscious of His Omnipotence. "True, we cannot see God; but from that can we say that He does not exist? We see multitudes of stars at night; we cannot see them by day. Does 'invisible' mean non-existent? We are finite and limited in our vision and so we are not able to see God. But for that reason let us not say, 'There is no God', or 'There is no need for Him!'"[4]

"The notion that God and Religion are superfluous and can be dispensed with are born out of a state of mind that is itself the product of age of God and religion. We are apt to be unconscious of what we possess, thanks to our parents and to our society. The sons of a wealthy father may imagine that poverty and unemployment are no problems, and that they are just vices and aberrations, born of idleness. Similarly, we who have inherited certain habits of thought and moral feeling are apt to think that God is a superfluity that can be dispensed with."[5] Rajaji declares that only faith in God and our devotion to Him can sustain us in life.

The acceptance of the idea of God, according to Rajaji, is not utilitarian in concept. The most enlightened scientists of the twentieth century have recognised the reality of the Supreme Being, though we do not know His form. One thing can be asserted, however, God is not like anything that exists or does not exist but which we conjure up in our imagination in our own imperfect way. "The worship of God," concludes Rajaji, "is an elemental necessity of life as it has evolved in man and his mind. It is on a par with the urge for the maintenance of the race and the urge for the preservation of life, like hunger and thirst, and the urge of sex. It is a fundamental of existence and survival, not what can be questioned on the basis of use and utility. Man's survival calls for detachment and therefore for God."[6]

Rajaji's study of the great religions of the world confirms his faith in God. He considers that all religions are but paths to the realisation of God and that the *Isvara* of the Hindus, the *Allah* of the Muslims, and the *Jehovah* of the Christians are all one and the same. They are different names given to the Divine Being by their followers. "Hindu religion like other religions is the worship of God. The special characteristics of Hindu religion are that it respects other religions and holds that worship offered by people following other religions through whatever" forms of words or rites bears good fruit just as worship in accordance with Hindu religion."[7]

The great historian, Dr. Arnold Toynbee, observes that Hinduism is unique among the historic religions of the world in its breadth of view that Truth is not the exclusive interpretation of Hinduism or any other religion. In the twentieth century with its great achievements in science, that seem more to be a menace to humanity than a blessing, men may adopt the doctrine of Hinduism that all paths lead to God, viewing it from an utilitarian angle of thought. Even this outlook on the need for accepting the existence of the Omnipotent Being is right teaching and gives us a true vision of reality. "In the Hindu view, as Dr. Toynbee has rightly emphasised, everyone of the religions of the world is a true vision and a right way. Each has a special spiritual value for the people who have developed it. To know this is good but it is not enough, says Dr. Toynbee. Religion is not just a matter for study, as Hindu scholars are apt to think. Truth is something that has to be experienced and to be lived."[8]

Rajaji was a devout follower of Hinduism. He made a deep study of its philosophy as contained in the *Vedas* and the and *Upanishads*, the *Bhagavad Gita* and other classics of Hindu thought as interpreted by saints and scholars down the ages like Sankaracharya and Ramanujacharya as well as realised souls like Gautama Buddha and Sri Ramakṛishna Paramahamsa, Swami Vivekananda and Sri Aurobindo, Ramana Maharishi and Mahatma Gandhi. Hindu religion is based on a ceaseless quest for Truth reared on a solid foundation of belief and faith in the existence of a Supreme

Being. It has in it a rare blend of philosophy and practical life that makes it an incomparable heritage for its adherents. "The Hindu doctrine of all ways leading to God with the resulting attitude of the Hindu creed towards all other religions, is unique. No other creed has arrived at this all-embracing fraternity of faiths, which Hinduism stands for. Judaism, Christianity and Islam do not give room for this doctrine of universality. In religious thought this universal validity is as great a discovery of truth as the law of gravitation in the realm of physics. Let us not give up this precious possession in a desire to copy others who claim exclusive possession of truth. Even in its conception of the Supreme Being, Hinduism differs from other monotheistic creeds in that Hindu philosophy places, the Supreme Being above the rule not only of Space and Time, but also of numbers. He is one and He is innumerable, Asankeyah, not subject to the law of numbers."[9]

The *Upanishads* are the quintessence of all religions in which is expressed the eternal yearning of the human soul for realizing immortality after a cycle of numberless births and deaths. The cry of men all along, as conveyed in the *Upanishads*, has been:

> Lead me from the unreal to the Real,
> Lead me from darkness to Light,
> Lead me from death to Immortality!

"The principal teaching of all the *Upanishads* is this: Man cannot achieve happiness through mere physical enjoyment obtained through wealth or through other goods of the world, or even through the pleasures attainable by elevation to the happy realms above through performance of sacrifices that are prescribed in the *Vedas*. The only happiness worth a wise man's seeking is permanent happiness as distinguished from fleeting pleasures that are exhausted by enjoyment like a credit account in a bank either here or in the world beyond... *Jehovah, Allah* and the God of the New Testament may well be made the central name piece of the teaching of the *Upanishads*, and the sense of it would remain unaltered. Pious men of all religions should indeed study the *Upanishads* and the *Gita* in that very manner, to whatever faith they may belong, only substituting their accustomed name wherever the Supreme Being is referred to."[10]

To Rajaji God is a reality, whether He is seen or unseen, whether men have belief in His existence or deny Him. The poet Wordsworth called duty or *dharma* "the stern daughter of the Voice of God." Rajaji avoids this intermediary between Man and his Maker, giving no importance to the role. When a person is hesitating to do some wrong act, an inner voice, generally called conscience, comes between the wicked thought and its fruition. Conscience is the whisper of God that prevents a person from straying from the right path and should be listened to as against the call of the baser self in man. God is necessary for life, by whatever name He may be called; He is within all men like the blood in their veins. "Call Him by the name of Conscience if you like," Rajaji says "but He is there speaking from your heart in spite of yourself. Conscience is not a phenomenon of schizophrenia or split personality but is God whispering to us. Let us listen to Him as Socrates did in Athens and as Gandhiji did in our country, in our own age. Listen to the Deity in your heart who speaks to you when you need Him. Appeal to the God in your brother's heart when you wish him to change his ways."[11] Rajaji concludes by saying that not only is God always with us but if we do not shut our doors against Him, He will keep us company always and speak to us if we do not plug our ears against Him.

Then comes the question in the mind of every human being: "What is the form of God?" Rajaji quotes from the Hindu Scriptures that God is unseeable and has also a clearly seen form. He is the infinite material Universe itself. His form is nothing like anything else known to man or unknown. Modern scientists can only contemplate God in terms of a hypothesis, but God cannot be the subject of a mere hypothesis, asserts Rajaji. On the other hand, the history of civilization down the centuries conclusively proves that God is the source of all morality, the bedrock of all spirituality, the basis of all the culture, which is the true wealth of mankind [12] "The First Universal Cause *Ishvara, Allah*, God, as it is variously called, has transformed itself into this Universe that we see. The Universe that our senses perceive is therefore an *avatar* of *Ishwara, Allah* or God whom the sages of all religions ardently desired to see but deplored they could not. The paradox is put in the shape of two names given to the Lord in our *Sahasranama*: *Adrisya* and *Vyakta Roopascha* in the same verse, one following the other, unseeable and also clearly perceivable."[13]

The aim of all men is not mere material prosperity or enjoyment born of sensual pleasures but something higher. After all is said and done, Man, being the monarch of all creation, should aim at something nobler than sheer animal existence summed up in living without any objective and dying without any achievement worth the name. There is a purpose in life beyond worldly advancement, namely, the realisation of spiritual wealth, which is incomparable with anything man can acquire. There is a denouement in death, once a man has worked out his *karma* and has shed all the earth in him and become a pure soul, a part of the Eternal Spirit pervading the Universe. This exalted peak of perfection can be reached only by devotion. "Truth and devotion are the only means to reach God. If we forget God, we shall die. Without devotion we cannot truly live. For devotion and Truth together make up wisdom. To bring God into one's own heart needs no great learning. Devotion will suffice. Devotion is the root of all religions. If there is real devotion in a person, his religion will serve his needs. Without devotion one cannot attain wisdom through religion or through conversion to other religions."[14]

Rajaji is catholic in his religious outlook and feels that all faiths are equally effective for man's realisation of his supreme goal in life. The quest for God is like the thirst of a wanderer in a desert; he must find an oasis and quench his drouth, otherwise he will perish. Hence those who yearn for a glimpse of God will not search for some new religion to satisfy their soul. A man whose earnest desire for God is overwhelming will seek the Omnipotent by following his own religion. "If we believe that God is everywhere," sums up Rajaji, "why should we not believe that He is in objects to which so much concentrated devotion is attached? Christians believe in the doctrine of original sin. Hindus believe in the doctrine of acquired sin. But all of us believe that God, by whatever name we call Him, is in the world. When we say that all ways lead to God, we do not mean only that we respect other religions. It is also implicit in that statement that most people find their way to Him most easily along the path, which they are taught to tread as children. You know that the English Christian, C.F. Andrews, came through his association with Gandhiji, to accept the fundamentals of our faith: yet when he died, his outlook had changed again, and Christianity appealed to him more than anything else. Similarly, among our sects, it is the Vaishnavite symbols that appeal to me most."[15]

Rajaji mentions that the two great source books of Hindu Religion as embodied in the doctrines of *Vedanta* are the *Upanishads* and the *Bhagavad Gita*. He compares the modern methods of scientific enquiry and investigation to the insistence on Truth and a ceaseless quest for the same as expounded in the *Upanishads*. The Sovereignty of the Divine is explained in the *Bhagavad Gita* in a manner that overcomes the questions of modern science when pitted against the problem of religious cosmology. The *Bhagavad Gita* declares that Divine Sovereignty is exercised through the inexorable laws of nature. "A study of the *Upanishads* will show that the *Vedanta* postulates that the Universe is the result of a gradual unfolding of the creative power inherent in the primordial substance. In fact it may be said that the philosophy of Hinduism anticipated the basic theories of biology and physics."[16]

A study of the Hindu religion and its philosophy convinced Rajaji that the need of the present times is *Vedanta*. For the foundation of *Vedanta* is a feeling of oneness with all humanity, its welfare and happiness. "The Vedantin is a citizen of the world," asserts Rajaji, "and a soldier in the world's army in a totally non-martial but no less heroic war against evil, the more heroic since he seeks no personal reward."[17] Those who have sought real enlightenment in the realms of the spirit have been most bewildered by the beauty of the Universe, far beyond the achievements of twentieth century science. They find no satisfaction in following faiths that have stopped further progress after passing a certain limit of intelligence and illuminating the soul. "The mysticism involved in the *Vedanta* relates the good life to truth and science. The conflict between religion and science is replaced and healed by harmony and integrated thought... *Vedanta* has a contribution to make to an enduring civilization. No polity based entirely on exploitation or force, even though able and well-intentioned men administer it, can last or be elevating even during the period it lasts. *Vedanta* offers a religious faith that can have no quarrel with the scientists who work in the laboratory or with the geologists, who do research in the history of the physical world, and yet it offers a firm spiritual foundation for the just polity of a new world."[18]

Rajaji sums up his supreme faith in God and his own philosophy of life as to the working of the Universe by its Great Creator in language that mirrors the ecstasy of his soul at the contemplation of the Divine. "The Supreme Power has arranged things so that they go on by themselves. He does not intervene. He has transferred all power to the things themselves. It is a grand and perfect scheme of decentralization. He is ever working and yet He is not seen to be working. The machinery works by itself once it started. We that observe and are curious, sceptical or credulous, we too are all parts of this Great Machine. We cannot get out of it and look through the windows into the Engineers' residence or workshop. This is the predicament of man, who is himself the most wonderful part of the Universe. It is not surprising, therefore, that some men say the machine is all automatic and there is nothing else, that there is no Engineer. The very smoothness and efficiency of the working of the machine leads to this notion."[19] Only a man of Rajaji's spiritual attainment could soar to such a splendorous vista of the Unseen Architect of the Universe!

Prayer played a great part in Rajaji's life, as is often alluded to by him in his writings. We are all God's children, he says, and we must approach Him as a child approaches its mother. It is said that we should be God fearing. No doubt it is true. But God-fearing does not mean that we should fear the Almighty just as we are afraid of a tyrant. The simple idea behind the phrase "God-fearing," declares Rajaji, is that we should not break the commandments of religion and be afraid of the consequences that it brings in its wake. If we approach God in all humility and lay bare our hearts to Him as a child does to its father or mother, He will come to our doors and keep us company. But if we close our ears, we cannot listen to His whisper.

It is possible for a man to have a vision of God if he keeps his heart pure by meditation. If it is polluted by lust or greed, God cannot be seen. "Our hearts are like mirrors. If they are pure and clean, they will help us to see God. Greed and anger are not easily got rid of. If we cannot quite get rid of them, we must turn them in a direction where they will not be so mischievous. Turn your desires and power into the quest for God. Turn even your anger towards God ! We tame wild animals. We bring under control even an elephant and make it obey us. Our mind, too, can be tamed and controlled if we make the effort. Wisdom is the god with which we can bring the mind under control."[20]

Just as he insists on the importance of *dharma* in a man's life the necessity for fulfilling its injunctions, he feels that devotion to God or *Bhakti* is the *sine qua non* for realizing the Supreme. A man need not be learned in order to instill faith in God into his life. His devotion to God will lead him to heaven. "Truth and devotion are the only means to reach God. If we forget God, we shall die. Without devotion we cannot truly live. Devotion is the root of all religions. If there is real devotion in a person, his religion will serve his needs. Without devotion one cannot attain wisdom through learning or through conversion to other faiths."[21]

An assessment of Rajaji's personality will not be complete without a reference to his reiteration about the need for men to render selfless service to others. "Of what use is a man's existence if he does not devote himself to the service of his fellowmen?" His whole life was a saga of service and sacrifice for his country and his countrymen. He laid his rich talents at their feet and laboured for the lowliest and the lost. From a freedom fighter against the British rulers of India, he transformed himself, after it had attained independence, into a fighter for securing the fruits of freedom to the masses, denied by those in power. Authority did not daunt him, power did not corrupt him, for *dharma* was his lodestar and truth his shield invulnerable! He may not have achieved all that he aspired for. But he was always a *Stitha-pragna* and could say in all humility: "I am not a sage nor a saint. I am a humble believer and that is all, and I wish to speak the Truth. I speak about my personal experience. Whenever I have keenly felt the distress of others and I pray for their relief, I have found God has answered. I say to others confidently, pray for others, and God will surely believe their distress and take care of you without your asking for it. If our concern is truly about other people and not about anything for oneself, God will help."[22]

Rajaji was not a mere professional politician who sought to acquire power and enjoy the plums of office. One of the exalted ideals of his interesting life was the interpretation of Indian culture. His deep study of Sanskrit and its classical literature like the *Vedas* and the *Upanishads*, the *Bhagavad Gita* and other books of Hindu religion and philosophy, the *Ramayana* and the *Mahabharata* enabled him to imbibe the essence of the wisdom and culture of ancient India. A scholar of English literature who delighted in the dramas of Shakespeare and the political philosophy of Bacon and Burke as well as modern writers and savants, he could speak and write with authority on the varied aspects of life. Gifted with a mind of encyclopaedic range, he gave to the world a rare feast of information and knowledge clothed in felicitous language, that shows what a remarkable mastery of thought and expression he possessed in his public life of over half a century of action and achievement. His writings on all subjects from the *Bhagavad Gita* to birth control are interspersed with illuminating passages that interpret Indian culture and make fascinating reading. They abound with quotations from Kalidas and Keats, Valmiki and Vyasa, Thoreau and Tiruvalluvar, Shakespeare and Shankara, Mahatma Gandhi and Ramana Maharishi, giving ample proof of his catholic taste in ancient and modern literature of all countries. His books and lectures serve as an education in themselves and the reader is presented with a panorama of life and thought down the centuries, giving us glimpses of his extraordinary mind and the depths of his profound passion for delving into the mystery of the Unknown.

It may be difficult to define the word "culture" but the culture of a nation is generally understood to consist in certain comprehensive aspects of their life and activities that reveal their contribution to the literary, artistic and spiritual wealth of mankind. The culture of a nation consists in the ideals that have inspired its people to immortal action down the ages and is deemed to be made up of its dreams and its deeds, their adventures in quest of the Supreme and for a true knowledge of the Divine. A country in the course of its history may be invaded by hosts of barbarians often but its glory and greatness in the fields of human effort and experience, its culture and civilization will live forever. Among the most renowned nations of ancient times India has survived with its civilization and culture untouched by time, unaffected by the vicissitudes of violent onslaughts on its way of life and thought, with its eyes unperturbed and serene ever turned towards the Eternal!

What amounts to culture is borrowed by Rajaji from the Kural of Tiruvalluvar, one of the classics of Tamil Literature, whose author was one of the wisest men of all time, comparable to Socrates and Marcus Aurelius, Confucius and Buddha. His definition of culture as delineated by a man of culture is as simple as it is sublime:

"Do not be mean; be noble, be large-hearted. Do not be false; be honest always and in everything; detest dishonesty, keep away from it as from offal. Do not be cruel; show tenderness towards every being that is weaker than you. Every loving thing loves its own life and is sensitive to pain as you are. Every being, be it child or girl or dog, or man or woman is sensitive to pain; be tender to them all. If you are mean, false or cruel, everyone will know it and know you are not a cultured person. Do unto others as you would be done by. This is the essence of culture in thought and action."[23]

Even as Mahatma Gandhi addressed prayer meetings to large masses of his own people almost every day of his later life, Rajaji gave lectures on the life and thought of India down the ages. His wide and varied reading of the ancient literatures of Sanskrit and Tamil and the modern writers in English from Shakespeare down to the present day enabled him to gather wisdom and disseminate it in unstinted measure to men and women who came to hear him or read his articles in the press or peruse his books. He made it a part of the mission of his life to bring to the knowledge of the common man the rare treasures of thought and radiant philosophy left by the ancient sages and *rishis* of India, in order that the masses may realise that they are the inheritors of the immortal legacy which they should not tarnish but render it more glorious and resplendent. Rajaji considered that this work on his part was more important than the role he had played under Gandhiji's leadership in bringing India to the threshold of freedom. According to him:

"The best service I have rendered to my people is the retelling of the *Ramayana* and the *Mahabharata*. This has been a source of great joy to me in the evening of my life. It is good to be a political and national worker and to take office and work hard. But I have seen that it is better to be able to leave it and enjoy the company of the sages of our land and to help them to speak to our men and women again."[24]

With this objective Rajaji spent a great deal of his time and energy, away from the din and tumult of politics, in rendering into English the great Indian epics, the *Ramayana* and the *Mahabharata*. A great scholar in Tamil and a profound admirer of Tiruvalluvar, he translated the *Kural* into English and revealed the wealth of its wisdom to the present day readers. His desire to interpret the wisdom of the world knew no bounds of creed or clime. He translated the works of Marcus Aurelius into Tamil and showed his countrymen the richness of mind of that remarkable Roman Emperor. Sankaracharya's *Bhaja Govindam* considered by Rajaji as containing the quintessence of his thought was rendered into English as forming part of the philosophy of India. Paying a tribute to the unknown author of the *Bhagavat Puranam*, he marvelled at the humility a man who left a gem of rare quality to the world but kept himself in the dark as to his identity or importance in the age to which he belonged. "Humility is an essential part of true culture," he remarks, "as mentioned in the *Gita*."

Rajaji's *Mahabharata* in English is not a mere translation of that classic, which depicts lofty delineation of character and deeds of heroism of men and women different from the ordinary mould of humanity. He says:

"*My Mahabharata* is not a translation but a condensation. Is it any wonder, after I have lived so long, that when I go into a jungle, I can cull the most beautiful flowers? I am fortunate to have been given the time to be able to do it. The real value of many of the stories is that they show us that great and good men can do bad things and that even the best of men cannot be virtuous all the time."[25]

Rajaji considers that the *Mahabharata* belongs to the world and not only to India. It is a classic of world literature like the *Bible* and the *Bhagavad Gita* the work of Valmiki and Vyasa, Kalidasa and Kamban, Shakespeare and Milton. Like other classics, the *Mahabharata* gives us unalloyed pleasure for all time, elevating our souls above the trivialities of earthly pursuits and the trifles of daily existence, and whispering into our ears, "the unbroken Voice of God, the Eternal Watchman and Friend". And to the Indian people down the ages it is a perennial fountain of spiritual strength. "The characters in the epic move with the vitality of real life ... The *Mahabharata* discloses a rich civilization and a highly evolved society which, though of an older world, strangely resembles the India of our times, with the same values and ideals ... The *Mahabharata* moulded the character and civilization of one of the most numerous of the world's people. How did it fulfil how is it still continuing to fulfil this function? By its gospel of *dharma*, which like a golden thread runs through all the complex movements in the epic; by its lesson that hatred breeds hatred, that covetousness and violence lead inevitably to ruin, that the only real conquest is in the battle against man's lower nature."[26]

The men and women represented in the *Mahabharata* makes the reader feel that he may meet them in real life as one walks in the streets of twentieth century India. We meet in the immortal epic the venerable Bhishma, the perfect knight, who gave up his right to a crown in order to please his father and make him happy. Therein is depicted Dronacharaya, the great master of archery who trained Duryodhana and his wicked brothers, having a soft corner for his only son Aswaththama and whose death was brought about by a subterfuge. Karna, the eldest son of Kunti, reveals his noble character by his rare fidelity to his masters, the Kauravas, refusing his mother's appeal to espouse the cause of his own brothers. Duryodhana, whose soul is inspired by a devilish desire to destroy the Pandava brothers, meets a well-deserted end at the hands of the mighty Bhima. The five Pandava brothers have become models of integrity and intrepid bravery on the battlefield—Yudhisthira, the matchless devotee of *dharma*; Bhima with the strength of a giant who destroyed the vile Keechaka for his foul design against

Draupadi's virtue; Arjuna the peerless warrior whose prowess on the field of battle was never surpassed by any hero in the legend or history of India; Kunti, the blessed mother of five sons who brought her glory and renown; Draupadi, who braved the brutal attempt to dishonour her in Duryodhana's Court by her dauntless faith in Lord Sri Krishna and Sri Krishna Himself giving to Arjuna and the world His immortal message in the *Bhagavad Gita*. Where else in the world's literature can we find such examples of extraordinary men and women, far from the multitude and towering above all mankind like Himalayan peaks? Rajaji sums up the greatness of the epic in these words:

"Learnt at the mother's knee with reverence and love, the *Mahabharata* has inspired great men to heroic deeds and also enabled the humble to face their trials with fortitude and faith. It is a noble poem possessing in a supreme degree the characteristics of a true epic, great and fateful movement, heroic characters and stately diction. One can even read a translation and feel the overwhelming power and incomparable vastness and sublimity of the poem. If a foreigner reads this book-translation and epitome though it is and closes it with a feeling that he has read a good and elevating work, he may be confident that he has grasped the spirit of India and can understand her people high and low, rich and poor."[27]

Rajaji had few peers among intellectuals of present day India though he spent his life in the political field having joined the freedom fight at the imperious call of Mahatma Gandhi. His knowledge of Sanskrit and Tamil was equal to his mastery over English. He translated the classics of Indian thought and wisdom, bringing enlightenment and delight to millions of readers. His original writing in Tamil was his outstanding contribution to its modern literature. His parables and short stories in Tamil contain the essence of his wisdom and experiences garnered from his wide study of world literature from Marcus Aurelius to Machiavelli, John Bunyan to Julian Huxley, Radhakrishnan to Bertrand Russell. Rajaji had as profound a reverence for the *Bible* as he had for the *Bhagavad Gita*, considering them both as containing the quintessence of the world's philosophy. He was imbued with the wisdom of Socrates as well as that of Sankara and admired the political acumen of Francis Bacon and Edmund Burke. He admired the consummate art of Shakespeare and delighted in his dramas considering the Bard of Avon supreme, in his sphere. His taste in literature was truly catholic and he sought to interest his countrymen in the perennial treasures of the philosophy of the prophets of mankind the unknown seers of the *Vedas* and the *Upanishads*, Gautama Buddha and Jesus Christ, Zoroaster and Mohamed, whom he deemed to be divine messengers sent down to earth to deliver the message of God to men.

The essence of Indian culture, in the estimation of Rajaji, is its exaltation of the ethical side of human existence focusing the end and aim of man's life in the quest for the realisation of the Supreme. A nation's progress lies in its ability to enthrone moral values among the masses as well as the classes. Thousands of years ago the *rishis* and sages of India laid stress on a man's spiritual endeavours as the *summum bonum* of his life as against the acquisition of material wealth. The survival of man depends on his faith in the supernatural governance of life and the acceptance of certain cardinal principles of probity and virtue. "We can no more survive as a species without this faith than we can dispense with the atmosphere of the planet, and hope to survive. The faith and the values emanating from that faith are part of our evolutionary history as the oxygen in the atmosphere of our physical life. We cannot dispense with faith unless we are prepared to destroy ourselves or become quite a different kind of living beings."[28] The advance of civilization depends on the acceptance of the existence of God and the acknowledgement of His Supreme authority. The Hindu scriptures have described God as unseeable and also as a clearly seen form. We see the form of God in the infinite material universe itself, if proof were needed of His existence. We cannot afford to give up this faith in God, for civilization rests on it. Mankind has grown up by climbing this ladder of light and cannot kick it down, lest its disappearance from our midst plunges us in illimitable darkness and despair. God may only be a hypothesis to the modern scientist but the civilization of centuries has shown that God is the living seed of moral values. Civilization is not only material progress but is the prevalence of respect for moral values. These values are man's special inheritance, without which advance in influence or power is not civilization."[29]

The insistence on the implementation of *dharma* is the keystone of Indian culture. The fulfillment of *dharma* is the foundation of all success in life and in the effort to reach man's divinely destined end. "*Yato dharmah, thatho jayah*!: Where there is righteousness, there is victory!" says the ancient Indian scripture. The two great Indian epics, the *Ramayana* and the *Mahabharata*, teach us that where there is a fall from *dharma*, ruin quickly pursues the man pitilessly to his eternal doom, whether it was Ravana or Duryodhana. *Dharma* rules the world and the lives of men whether they believe in it or deny it, for *dharma* is nature. Just as man is ruled by *dharma*, Government also has to accept the superiority of *dharma* in exercising their authority over those under their rule ... Government floats in a sea of *dharma*. Without *dharma* Government is a sham and an evil, an infliction and a danger instead of being a protection."[30] All men and women may not put *dharma* into practice in their daily life. But they must have reverence for *dharma* and be afraid of breaking its injunctions, although they may fail in their efforts to follow it. For, the fulfilment of *dharma* by individuals will raise the moral level of the nation as a whole and evoke respect for its character from the whole world. "According to India's *dharma*, all the wealth, talents and good luck of which an individual happens to be the possessor, come from God and he holds them as a trustee for the benefit of those around him. Giving is the biggest employment one can extract out of one's possessions. There should be no legal compulsion intervening in the working of this moral trusteeship. Giving ceases to be enjoyment if it is subjected to coercion."[31]

Rajaji considers that the genius of India is her spirit of toleration, which she has exhibited right down her history. That is also the secret of the survival of Indian culture and civilization for the past five thousand years or more while other countries and civilizations have been destroyed by the tornado of time. The culture of Greece is a mere memory, the glory that was Rome is a relic of the past, the mighty empire of the Moors in the Middle Ages is but a chapter in the history of mankind. Only India has withstood the ire of her conquerors and their scant respect for things of art and beauty, worshipping the spirit and turning her back on all that is transient and ephemeral, ever in quest of the Eternal! This is because her moral glory has overcome all her hostile invaders who ruled over her but were never able to hypnotize her mind or conquer her soul. They have been absorbed in her all embracing arms of toleration becoming part and parcel of her innumerable family, growing in diversity with the passing of time. "Hinduism is the only religion that specifically includes in its doctrine that all ways lead to God. This catholicity in Hinduism is unique. Other religions do not seem to go so far as does Hinduism in this respect."[32] Dr. Arnold Toynbee observes that Hinduism is unlike any other world religion in holding that truth is not the exclusive representation of Hinduism. "In the Hindu view every one of the religions of the world is a true vision and a right way. Each has a special spiritual value of its own for those people who have developed it. To know this is good, says Dr. Toynbee, but it is not enough. Religion is not just a matter of study, as Hindu scholars are apt to think. Truth is something that has to be experienced and to be lived, and this is the field in which Sri Ramakrishna has manifested the uniqueness of Hindu thought."[33]

The essence of the *Bhagavad Gita* is explained by Rajaji in easy language which enables an ordinary man to elucidate its meaning and fulfil its injunctions in everyday life. This great classic of Hinduism has been the subject of erudite commentaries by men like Sankaracharya and Ramanujacharya in the past, Tilak and Aurobindo at the present day. Mahatma Gandhi, among his multifarious activities, found time to give his own interpretation of its profound thought and practical wisdom. Rajaji explains the essence of the *Bhagavad Gita* as follows:

"The *Upanishads* and the *Gita* contain the way of life which, by tradition has been accepted by the people of India. What does the *Gita* teach us? Every moment our thoughts and actions shape our future. As a result of each thought, which we entertain and dwell on and each act we commit, we become pure or impure, good or bad, more and more restrained and capable of self-control or less and less so and slaves of evil urges. Every moment we shape our character as a goldsmith shapes his jewels on the anvil, each stroke of his hammer corresponding to each thought, word and act of ours. Therefore, it follows that by self-control, enlightened thought and action we can free our souls from evil and improve ourselves ... The *Gita* tells us how to secure tranquility in the midst of all work. Tranquility cannot be secured by taking *Sanyas*[34]. It can be secured only by detachment and by placing our trust in the Lord in all things. Do your work not thinking of success or failure or what results are to be reaped by you by that work. Throw the burden of results on the Lord when you do the work. This is the way to keep the soul clean whatever may be the work one may have to do. The butcher in the *Mahabharata* is a famous example of this teaching."[35]

It was Swami Vivekananda who first opened the eyes of the West to the enlightened teaching of the *Vedanta*. He created a sensation when he first appeared at the Parliament of Religions in Chicago in 1893. His ochre robes and his outstanding personality became the cynosure of all eyes in that distinguished gathering and he conquered the West with his interpretation of India's ancient wisdom and philosophy. It proved to be the beginning of a brilliant saga of a spiritual invasion of America and Europe by the "Hindu Monk from India". It has been followed since then in more and more centres being opened there for the spread of the culture and civilization of India and the creation of a yearning in Western minds for understanding the soul of India and exploring the unknown realms of its renascent religion, Hinduism. A great admirer of Swami Vivekananda, Rajaji expounds the philosophy of *Vedanta* in a direct and forceful manner, reminiscent of the most renowned disciple of Sri Ramakrishna Paramahamsa. Rajaji poses the question and answers it himself: "Is it possible at this stage of human history to build a religious fabric around scientific truth as it has evolved and is still evolving? *Vedanta* is the answer. It is not necessary to build a new religion. In India we have a religion and a philosophy attached to it, as old as civilization itself which is remarkably consistent with science as well as politics.... Put in precise words the claim is that a code of ethics and a system of values were evolved by the Hindu philosophers out of the religious philosophy known as *Vedanta*, which is not only consistent with science, but is also admirably suited to be a spiritual basis for the most just and stable social organization that good people all over the world desire and are working for... It goes without saying that spiritual values proposed as the basis of a sounder social organization must not be an improvisation or an invention of expediency designed to further material interest by cloaking it with sanctity. A spurious scheme of the so-called spiritual values to serve a sordid purpose would be a delusion if self-imposed, an imposture if offered for acceptance. Honesty is the best policy; but it is not as policy that honest conduct was made part or continues to be part of every religion. Similarly, *Vedanta* is bound to help regulated economy but it is not for that reason that it was conceived or should be accepted as a faith...

"The *Upanishads* and the *Bhagavad Gita* are the sourcebooks of *Vedanta*. It is a remarkable achievement of intellectual imagination it would not be correct to call it inspiration that the rule of law in science was anticipated in the ancient Hindi scriptures. The God of *Vedanta* is not an anthromorphic creation with human capriciousness and desire for power—a conception against which the veriest tyro in modern science can launch a successful attack. Divine Sovereignty is explained in the *Bhagavad Gita* in a language, which anticipates and meets the difficulties that modern science raises against religious cosmology. According to the *Bhagavad Gita*, the Sovereignty of God is exercised in and through the unchangeable law of cause and effect, that is, through what we call the laws of nature...

"All culture in India has been rooted in *Vedanta*. Whatever courage, heroism, self-sacrifice or greatness is to be found in our history or seen in the lives of our people has sprung from *Vedanta*, which is in our blood and tradition. It is in the air, so to say, of India and of Asia. The foreigner has to get it from books and he necessarily sees so much subtlety in it that he may well swear that it is impossible that such a doctrine could ever be the actual cultural basis or living spiritual principle of the daily life of any people of modern times. For, *Vedanta* is undoubtedly a living philosophy of life in India, a part of the mental structure of our people. The people of India get it not from a study of books but from tradition. The greatness of Gandhiji and the strength of his movement are entirely derived from and rooted in *Vedanta*. However much foreign civilization and new aspirations might have affected the people of India, this spiritual nutriment has not dried up or decayed or changed. The lives of the rich as well as of the poor, of the leisured classes as well as of the peasants and labourers, of the illiterate and not only of the learned, are in varying measure sustained and sweetened by the pervasive fragrance of this Indian philosophy. Paradoxical as it may seem, even communities deemed dishonest and disreputable have evolved a code of honour of their own, and are Vedantins to the extent of sincerely respecting it. This curious moral enclave in sinful lives touches the heart and makes a great piety of what is doubtless just a matter for sheer reprobation."[36]

Rajaji reminds the reader that there is all over the world a wave of enthusiasm for regulating the work of the individual in the interests of society. There is therefore a need for creating a social conscience in men combined with a cultural desire to contribute to the common good of all by fulfilling one's *dharma* or duty leading to a spiritual quest. This cannot be achieved by the enactment of any number of Acts of Parliament. "The spiritual value of things depends on the price paid for them in terms, of suffering and sacrifice. An easily achieved revolution has not the same psychological virtue as one paid for in blood and tears. Where there is no backing of revolutionary fanaticism or its after effects, there must be found something else to operate as motive power. The only thing that can do this effectively is a faith that operates as a law from within and co-operates with State imposed restraints. In *Vedanta* we have a teaching rooted in immemorial tradition and associated with the sacred names and memories of a long line of seers, which can serve as the spiritual and cultural basis for a new and more just economy of life, if not all over the world, at least in India itself."[37]

"*Vedanta* is not mere philosophy." Sums up Rajaji. "It is both philosophy and religion. Yet there is no controversy in it about form and worship. *Vedanta* is the common heritage of the people of India whatever denomination they may happen to have been brought up in."[38]

Rajaji often gave a spiritual tone to his writings and made an appeal to men and women to follow *dharma*. Though he dealt with every day problems of political life in India and the world, he raised his ideas to a pinnacle of elevated thought. He based his philosophy of life on ethical principles that are eternal and the common legacy of all the great religions of the world. The result is that he elevates the reader to an atmosphere that inhibits all ideas of transient glory and impermanent achievement. He impresses upon people the power of prayer and the significant part it plays in the shaping of their daily lives for their good. "I have come more and more to believe in the efficacy of earnest prayer. I feel that all good men throughout the world should pray to the most high for the restoration among mankind of trust in one another's words and pledges. Most of the nations involved in this crisis are those who have accepted the message of Christ. They maintain great churches and cathedrals where they periodically reaffirm their homage to Christ. But the people of that vast country, Russia, do greatly still maintain their faith in Christ, though the government of Russia may have discarded Christ. Everyone now sees what will surely follow if Christ is discarded. It will be the same if Confucius or Mohamed or Buddha or Sri Krishna or Sri Rama goes. We want mutual trust again to rule the conduct of mankind as a species desiring to survive. We want to save ourselves from self-destruction. Be it Christ or any other Divine Messenger or Reverend Teacher that nations have accepted and honoured, we must bring all these divine leaders of men back into power if we do not wish to be doomed to destruction, and what is even worse living in unending fear until that day of annihilation."[39] It is Rajaji's firm belief that through which the world is passing can only solved for the good of mankind by going back to the good days when the prophets commanded reverence from the common man and their words were looked upon as a message from the Divine and followed by the people in their daily lives in a spirit of holy adoration and humble acceptance.

Rajaji was intensely proud of his country and his countrymen, its heritage of ancient culture and civilization, its emergence as the leader of Asiatic and African nations by its unparallelled feat of winning its freedom without the use of force or violence under the guidance of a man of God, Mahatma Gandhi. Again and again he reminds the reader how India has kept her soul radiant and serene accepting the challenges of chaos and confusion in the course of her history and triumphing over time and circumstance down the centuries, because of her ceaseless search for spiritual light amid the evanescent things of earth that tempted other nations and toppled them to their doom never to rise again! He attributes this supreme position of India in the world to her unswerving adherence to moral values and her quest for things of the spirit, turning her face away from the lure of the world and a transient life of low materialism. He often emphasises that eternal values should be enthroned in the hearts of men as worthy of their effort and endeavour in terms of the *Upanishadic* prayer. But his patriotism does not blind him to the faults and foibles of his countrymen that have cropped up after they attained their independence, namely, the corruption in the administration, the callousness of the rulers to the welfare of the common man, the suppression of moral values in an ape like pursuit of the material prosperity of Western nations. He poses the question why there should still be falsehood and dishonesty in the life and thought of the people and answers in telling words and true:

"Our country has attained freedom. We have a Republican form of Government. Our country holds an honourable place among nations... In spite of all this, why do we still behave like beggars and cheats? Why do we believe that falsehood and dishonesty are ways to prosperity? Why do we act like mean people who live on garbage? Without realising the greatness of *Bharata Mata*; why do we basely degrade ourselves? Let us spurn and leave behind the qualities belonging to our old bondage, and let us quickly assert our real, honourable nature. Poverty is nothing to be ashamed of but littleness of heart and mind is. We may live honourably in poverty. Let us not be bad men. May we cherish the new won glory of our motherland. Let us discard the mean, cowardly qualities of falsehood and dishonesty."[40]

REFERENCES

1. *Swarajya,* 3 January 1970.
2. *Bhavan's Journal*, 1 January 1969.
3. *Ibid.*
4. *Ibid.*, 16 December 1956.
5. *Ibid.*
6. *Ibid.*
7. *Swarajya*, 5 June 1971.
8. *Ibid.*
9. *Ibid.*, 27 November 1971.
10. Rajaji, *Hinduism: Doctrine and Way of Life by Rajaji* (Bombay: Bharatiya Vidya Bhavan), pp. 38, 41.
11. *Swarajya,* 4 September 1971.
12. *Ibid.*
13. *Bhavan's Journal*, Annual Number 1965.
14. Rajaji, *Ramakrishna Upanishad*, p. 85.
15. Felton, Monica, *I Meet Rajaji, op. cit.*, pp. 38-39.
16. *Swarajya*, 3 August 1968.
17. Rajaji, *Hinduism: Doctrine and Way of Life, op. cit.*, p. 94.
18. *Ibid.*
19. *Bhavan's Journal*, 6 August 1971.
20. Rajaji, *Ramakrishna Upanishad, op. cit.*, pp. 81-82.
21. *Ibid.*, p. 85.
22. *Swarajya*, 18 March 1972.
23. *Rajaji, Our Culture*, (Quotation from his translation of the Kural), pp. 7-8.
24. *Bhavan's Journal*, 7 January 1973.
25. Felton, Monica, *I Meet Rajaji, op. cit.*, pp. 38-39.
26. From the Introduction to The Mahabharata
27. *Ibid.*
28. *Bhavan's Journal*, 27 July 1969.
29. *Ibid.*
30. *Swarajya*, 23 March 1963.
31. *Ibid.*, 28 May 1966.
32. *Ibid.*, 15 November 1969.
33. *Ibid.*, 5 June 1971.
34. '*Sanyas*' means renunciation of the world.
35. *Bhavan's Journal*, 6 August 1961.
36. Rajaji, *Hinduism: Doctrine and Way of Life, op. cit.*, pp. 35-36.
37. *Ibid.*
38. *Ibid*, p. 40.
39. *Swarajya*, 3 August 1968.
40. *Bhavan's Journal*, 18 March 1972.

7

SOCIAL JUSTICE

Among the many causes dear to Rajaji's heart, which attracted his devotion were the uplift of the lowly and the lost, the friendless and forlorn. These found an important place in his thought. He was an upholder of social justice to the lower strata of Indian society, which, in spite of independence, were still perforce continuing in their abject plight of penury and want, the victims of the sons of Mammon, piteous pawns in the political game of the leaders of the land. The coming of freedom had not brought for them any appreciable improvement in their economic condition or enlightenment in their lives. They were still addicted to evils like drink that ruined them body and soul. They were still compelled to live in unhealthy surroundings and often fell a prey to disease. They were still poor specimens of humanity compared to the free people of other countries, although their devotion to *dharma* and their faith in God was higher than what prevailed among the elite of Indian society. It saddened Rajaji to think that the snapping of the bonds of subjection had brought no change for the better in their lives. They were now at the mercy of brown bureaucrats instead of their white masters before Gandhiji won liberty for them. The Father of the Nation had, no doubt, made them free but they had yet to enjoy the fruits of freedom like the richer classes. Mahatma Gandhi had called upon the people to give up drink, to banish untouchability from the Hindu fold and to dedicate themselves to the service of their fellowmen. "Those who serve the poor are great in the eyes of God," he had declared on a memorable occasion. But, alas! the country swarmed with self-seekers who were in power and authority. They made hay while the sun shone, more when the sun did not shine! Corruption stalked the land like an evil spirit, tempting even the most innocent with its prizes and rich spoils. Nepotism played an insidious role in raising men from the dust to roles of distinction. The nations of the world wondered if this was the land of Mahatma Gandhi, the apostle of truth and *ahimsa* even as his own people wondered if this was the *Rama Rajya* of Gandhiji's dreams! These aspects of life in India after *Swaraj* were ever-present before Rajaji's eyes and pained his innermost soul. He fought to mitigate those evils that had lowered India in the esteem of world nations and had led to the moral degeneration of his countrymen and their rulers.

Mahatma Gandhi considered that the eradication of the drink evil was one of the most urgent and effective steps necessary for the regeneration of the country and the economic prosperity of the masses. With that object he had included prohibition among the principal planks of his political programme for winning *Swaraj*. Drinking is interdicted by all the principal religions of the world. Hinduism, Islam, Christianity and Buddhism. Any true follower of these great faiths in this regard must implement their injuctions. Gandhiji was aware of the existence of this pernicious habit among the hamlet-dwellers as well as the members of high society to the detriment of their health and the depletion of their purse.

A realistic picture of the havoc wrought by drink on a poor man's home and family cannot be bettered for its grim truth than these words of Mahatma Gandhi:

"Only those women who have drunkards as their husbands know what havoc the drink evil works in homes that once were peaceful and orderly. Having identified myself with labour, I know what ruin drink has brought to the homes of labourers given to drink. Drink revenue makes people pay for their own corruption, moral, mental and physical. The loss of revenue is only apparent. The removal of this degrading tax enables the drinker, that is the taxpayer, to earn and spend more."

Rajaji was in complete agreement with his Master on the urgent necessity of ending this monstrous evil that has at times converted men into animals and incited them to commit the most heinous crimes including murder. When he became the First Chief Minister of

Madras in 1937, one of his earliest acts to make life better for the masses was the enactment of the Madras Prohibition Act. Rajaji applied it first to his own home district Salem.

"When I see the bazars in Madras City," wrote Rajaji in 1972, "studded with blazoned sign boards of licensed wine shops and young men in small groups encouraging one another visiting these shops, my eyes are filled with tears for what I have brought about without intending it, for I imagine I am not a little responsible for the DMK being in office. Our young men, hitherto totally innocent and ignorant of the allurements of brandy, whisky or wine, are being permanently sold to the Devil for the sake of a trifling addition to the Government revenue."[1]

The extra revenue derived from drink, argued Rajaji, was as immoral as it was undesirable, for it would be at the expense of the poor and their health and their happiness. The justification for the re-opening of toddy shops and arrack shops in Tamil Nadu after 20 years' closure on the ground that it would bring in more revenue to the government appeared to him to be puerile and opposed to all principles of wise statecraft and statesmanship. The welfare of the poor and the prospect of a decent, healthy existence for them necessitated the continuance of prohibition. "Drink revenue is a deceiver, even as the temporary stimulation that intoxicating drinks give is a deceiver. We do not want a revenue derived by opening the flood gates of wasteful, unhealthy and ruinous expenditure by the poorest families of the nation, who even now do not earn enough to meet the cost of normal healthy family life, owing to high prices, which cannot be held mechanically or by force but only by a great change in government policies. What will ruin the people cannot be good for state governments because just a little more revenue can be got out of it."[2]

What is morally wrong as well as against the interests of millions cannot be a right policy of the government. True that the reformation of the people is not in the hands of the State but government owes a duty to see that they do not put the drink in the minds of the people leading to that work havoc with their happiness in their homes and the harmony of their family life. "Public opinion in India among poor people is in favour of prohibition and will continue to be so in spite of the propaganda of the upper class whisky and brandy bibbers. These upper class drinkers can afford to pay for the liquor they consume and can fairly easily get special permits to waste their money and their health. They are for doing away with prohibition because they feel ashamed to ask for and obtain permits. The difficulty should be removed, they feel, if rich and poor, all were relieved from the Prohibition Law. The shame they feel is an index of the public opinion against drinking, and that makes all the difference between the American conditions, which brought about a repeal of prohibition there and the conditions prevailing in India. Prohibition is the one great reform that gives a chance to the poor people to go up in social status and in family comfort.[3]

The miserable life led by slum-dwellers in cities and large towns made Rajaji take up their cause. The rapid industrialisation of India during the twentieth century has attracted millions of people from the villages to the towns to eke out a better existence. This large influx of population has not been provided with adequate housing and other conveniences for a decent and comfortable life either by the government or by their employers. The result has been that though their economic condition has improved, the poorer classes of employees in mills and factories are forced to live in unhealthy and unsanitary conditions, devoid of sufficient amenities of civilized life.

Rajaji's heart went out in sympathy to these sons of the soil migrating from the villages only to land in slums. "The rural families migrate into the urban centres where they can get better wages, however miserable their living conditions have to be. There is no chance of getting rid of the slums... unless the matter is looked into as an occupational problem and not merely a question of replacing palm leaves and old newspapers by brick and cement at government expense.[4] He suggested that adequate work must be provided for the rural population through vast projects like road making, the building of tanks and reservoirs, and the dispersal of industrial centres to the countryside, so that they might earn adequate wages without crowding into the cities and being compelled to become slum-dwellers.

The starting of lotteries by the State in various parts of India, with tempting prizes running into lakhs of rupees, came in for severe condemnation at Rajaji's hands. Instead of making people work-minded and more industrious, these lotteries are a lure to laziness and hold men in their vicious grip. "It is wrong to get poor people's money by tempting them with intoxicating drugs through licensed agents of government; it is doubly wrong to do it by tempting them with the chance of winning a big prize, the odds being one against lakhs doubly wrong because the government does not give anything in return whatsoever to the lakhs of people who part with their money on this temptation and do not win the lottery, not even a drink. An act, which is morally wrong does not become right because the government wants money. Otherwise we could organize plain, forthright robbery on behalf of government instead of taxes sanctioned by legislatures. Getting money through the temptation involved in lotteries is not different from robbery, even as a man who cheats through a trick is no better than a thief ... I am convinced that it is evil to turn people's minds all over India to making money by luck and to tell them that if they do not succeed, their luck is at fault, and this is done just when the need of the hour is to stimulate hard work and divert men from the devastating propensity of shirking work and to look to cheating one another. The need of hour is to make men work minded. Lotteries do the site of what should be done."[5]

"Those who serve the poor are great in the eyes of God," declared Mahatma Gandhi who had dedicated himself to the uplift of the lowliest and the lost, living amidst them whenever he had an opportunity, sharing their joys and sorrows. Rajaji was moulded after his Master and always pleaded for the improvement of the lot of the poor. He called upon men and women to become interested in their welfare and observed: "Of what use is man's life if he does not serve those around him?" He condemned the Congress government for heaping too many burdens on the shoulders of the lower sections of the society in the shape of crushing taxation when they were hardly able to make both ends meet. He took up cudgels on behalf of the craftsmen of the goldsmith community when the Gold Nationalisation Bill was passed into law by Parliament, depriving millions of these artists of their legitimate profession and driving them to the doors of penury and want. The ostensible reason given by the government for the above enactment was to put a stop to smuggling of gold into India. "If you wish to stop the smuggling of gold into India and can do it, by all means, stop it," observed Rajaji referring to the Gold Control Bill. "It is indeed monumental stupidity to put millions of honest, skilled artisans out of occupation, because with all your extensive powers as government you have failed to stop somebody's criminal practices. If you cannot tackle officially defined crime, confess it but do not practise extensive and undeserved cruelty on innocent people to cover up your incapacity to govern. The smuggling is the automatic result of your depreciated currency, and it cannot be stopped by torturing the goldsmiths or any other particular community who have nothing to do with smuggling."[6]

"An official takes to corruption as a fish takes to water," observed Kautilya in his famous treatise on politics and statecraft, *The Arthasastra*. The prevalence of widespread corruption in the administration has been accepted on all hands and the attitude of the authorities seems to be to brush it aside since it is so common all the world over. Rajaji bewailed against the existence of this dark blot on the escutcheon of Indian administration. He recalled the great days when Gandhiji led the country to liberty, fighting on a high moral plane, fusing the two ineffable ideals of truth and non-violence into the invulnerable weapon of *Satyagraha*. He ascribed the fall in the standards of purity in the administration and the probity of its personnel to the belief of the Congress rulers that the end justified the means and that so long as success attended their efforts, the motives behind their victorious march need not be taken into consideration ! Another reason for the official classes deserting the path of *dharma* was the emphasis laid on making money somehow and enjoying themselves. "The widespread prevalence of corruption is recognised and treated as a malady whereas it is the result of trying to bring into being what calls for an honest profit incentive but without furnishing that condition. A philosophy of love of money is deliberately preached, but no room is given for making money in an honest way and the philosophy or psychology, whatever we call it, finds the way of corrupt practices... Production of wealth demands its toll, viz. the profit motive, if we dislike the way of slave-driving. Provided free enterprise is not only tolerated but encouraged and accepted as a decent and legitimate way of rapid and efficient industrialisation, and individual business is recognised and respected as a national enterprise and released from the numerous curbs and the impediments now imposed, and if foreign aid is channelled through non-State free enterprise, instead of through government, India and other undeveloped countries can be made to produce much satisfactory results in marvellously quick time. Instead, we seek to reach the goal through State Socialism, cutting out the profit motive, which is essential for sustained human activity. State Socialism is not compatible with free life. It must be accompanied by the compulsions of Communism or Fascism. Our friends who give lavish aid, and are disappointed with the results, blame Sankara and Buddha and the *Rishis*. I plead not guilty on their behalf, the failures are due to wrong policy, not due to our culture or the *Rishis*".[7]

Rajaji, however, did not throw up his hands in despair at the prevailing state of affairs in India. On the other hand, the canker of corruption and its concomitant evils could not be got rid of, he admitted, by refusing to confess the reality and coming silent victims. There was hope for the country if its rulers could shed their fondness for false standards of prosperity and progress aping Western countries, without copying the good in them or their two great virtues of duty towards discipline in their daily conduct. "We must elevate the simple life to the status it had enjoyed in Gandhian and pre-Independence days. It is 'standard of life' that has corrupted and is corrupting our souls. We should make our administration less expensive by reducing the number of people engaged in that unproductive but important work while, at the same time, paying adequate salaries to those employed. The development of productive industries should be unhampered by controls and by bureaucratic hurdles, so that they may grow quickly and absorb more of intelligent young men instead of their being driven by necessity to government service."[8]

The *sine qua non* for a satisfactory means of ending corruption permanently and restoring integrity into the administration lies in the inculcation of a high moral purpose into the minds of the official classes. They must be inspired with a feeling that the interests of the country demand their honest performance of their duties and that they must give their best, in spirit and in letter, for the betterment of their fellowmen and the fulfillment of the highest ideals of probity, efficiency and good government. Public officials must tread the way of righteousness, dedicated to the service of the country, realising that the path of duty is the way to glory. "Less taxation and less inflation, abandonment of the wholly wrong plan of finding industrial capital by oppressive taxation, and release of private capital and private initiative from the barbed wire entanglements of central planning, these will help to a large extent in clearing the air of the poisonous smog of corruption... Human nature cannot tolerate this state of affairs. The crisis will lead to revolution of some kind, communist or fascist or military. The people have to face this corruption on the one side, and high prices and unbearable tax levies on the other. Oppression and corruption must lead to revolt, and passing through anarchy, democracy must turn to dictatorship. Neither can the resulting tyranny escape the total evil of the times. The dictatorship will not be a relief for that, too, will be corrupt."[9] Rajaji concluded that only rectitude of conduct and righteousness of life would save the nation when faith in God was restored.

REFERENCES

1. *Swarajya,* 5 February 1972.
2. *Ibid.*, 29 April 1967.
3. *Ibid.*, 2 March 1968.
4. *Ibid.*, 3 July 1971.
5. *Ibid.*, 21 September 1968.
6. *Ibid.*, 2 January 1965.
7. *Ibid.*, 23 March 1963.
8. *Ibid.*, 4 May 1963.
9. *Ibid.*

BIBLIOGRAPHY

Ahluwalia, B.K. and Shashi Ahluwalia, *Rajaji and Gandhi* (New Delhi: Allora Publications), 1978.

Ashe, Geoffery, *Gandhi* (Bombay: Asia Publishing House), 1968.

Azad, Abul Kalam, *India Wins Freedom* (New Delhi: Orient Longmans), 1959.

Baliga, B.S., *Tanjore District Handbook* (Madras: Government of Madras), 1957.

Banerji, A.K. *Aspects of Indo British Economic Relations, 1858-1898* (Bombay: Oxford University Press), 1982.

Banerji, Arun, *Finances in the Early Raj: Investments and the External Sector* (New Delhi: Sage Publications), 1995.

Banerjea, S.N., *A Nation in Making* (London: Milford), 1925.

Bapat, S.V. (ed.), *Reminiscences of Lokamanaya Tilak* (Bangalore)

Besant Annie, *How India Wrought for Freedom: The Story of Indian National Congress* (New Delhi: Michiko Panjathan), 1975.

Bhattacharya, Buddhadeva (ed.), *Freedom Struggle and Anushilan Samiti* (Calcutta: Anushilan Samity), 1979.

Bimal Prasad (ed.), *A Revolutionary Quest* (Delhi: Oxford University Press), 1980.

Birkenhead, *Earl of Halifax* (London: Hamish Hamilton), 1965.

Birla, G.D., *In the Shadow of the Mahatma* (Bombay: Vakils, Feller and Simons), 1968.

Bose, Nirmal, *My Days with Gandhi* (Calcutta: A. Chatterjee), 1953.

Bose, Subhas, *The Indian Struggle* (Bombay: Asia Publishing House), 1964.

Brecher, Michael, *Nehru* (London: Oxford University Press), 1959.

Buch, M.A., *The Rise and Growth of Indian Liberalism* (Baroda: Baroda Press), 1938.

Chandra, B., M. Mukerjee, A. Mukerjee, K.N. Panikkar & S. Mahajan, *India's Struggle for Independence 1857-1947* (New Delhi: Penguin Books), 1989.

Chandra, Bipan, *Communalism in India* (New Delhi: Vani Educational Books), 1984.

Chandra, Bipin *et al.*, *India's Struggle for Independence, 1857-1947* (New Delhi: Penguin Books), 1989.

Chandra, Bipan, *Modern India* (New Delhi: NCERT), 1976.

Char, K.T. Narssimha, *C. Rajagopalachari: His Life and Mind* (New Delhi: Heritage Publishers), 1978.

Chatterjee, Bimanesh, *A Thousand Days with Rajaji* (New Delhi: Affiliated East-West Press Private Limited), 1973.

Chinatamani, C.Y., *Indian Politics Since Mutiny* (Allahabad: Kitabistan), 1947.

Chirol, Valentine, *India Old and New* (New Delhi: Lights and Life Publications), 1975.

Chirol, Valentine, *The Indian Unrest* (New Delhi: Lights and Life Publications), 1979.

Collins, Larry and Lapierre, Dominique, *Freedom at Midnight* (New Delhi: Vikas Publishing House), 1997.

Dalton, Dennis, *Indian Idea of Freedom* (New Delhi: Manohar), 1982.

Das, Durga, *Curzon to Nehru and After* (New York: John Dobu), 1970.

Dasgupta, Ajit K., *A History of Indian Economic Thought* (London: Routledge), 1993.

Datta, V.N., *Madan Lal Dhingra and the Revolutionary Movement* (New Delhi: Vikas Publishing House), 1978.

Dawarkadas, Kanji, *India's Fight for Freedom: An Eye Witness Story, 1913-37* (Bombay: Popular Prakashan), 1960.

Desai, A.R., *Social Background of Indian Nationalism* (Bombay: Popular Prakashan), 1976.

Desai, Mahadev, *The Story of Bardoli* (Ahmedabad: Navajivan Publishing House), 1929.

Doke, J.J., *Mahatma Gandhi* (Oxford: Oxford University Press), 1977.

Felton, Monica, *I Meet Rajaji* (London: Macmillan), 1962.

Gandhi, M.K., *An Autobiography or The Story of My Experiments with Truth* (Ahmedabad: Navajivan Publishing House), 1983.

Gandhi, M.K., *Hind Swaraj* (Ahmedabad: Navajivan Publishing House), 1983.

Gandhi, Rajmohan, *The Rajaji Story: A Warrior from the South* (Bombay: Bharathan), 1978.

Ghosh, P.C., *From Nagpur to Lahore* (Comillia: Abhoy Ashram), 1930.

Glendevon, Lord, *The Viceroy at Bay* (London: Collins), 1971.

Gopal, S. (ed.), *Selected Works of Jawaharlal Nehru*, Second Series, Vol. 4 (New Delhi: Jawaharlal Nehru Memorial Fund), 1986.

Gopal, S. (ed.), *Selected Works of Jawaharlal Nehru*, Second Series, Vol. 5 (New Delhi: Jawaharlal Nehru Memorial Fund), 1987.

Governor's Speeches (Calcutta: Government of West Bengal), 1948.

Guha, A.C., *First Spark of Revolution* (Delhi: Orient Longman), 1972.

Hardayal, *Hints for Self Culture* (Delhi: Rajkamal Publications), 1948.

Holmes, T.R., *History of Indian Mutiny* (London: Macmillan), 1898.

Irschick, *Political and Social Conflict in South India* (California: University of California Press), 1969.

Iyengar, A.S., *All Though the Gandhian Era* (Bombay: Hind Kitabs), 1950.

Iyengar, Masti Venkatesa, *Rajaji* (Bangalore: Jeevan Karyalaya), 1975.

Kakar, Sudhir, *The Colour of Violence* (Delhi: Viking), 1995.

Kapoor, A.N. and V.R Gupta, *A Dictionary of Gandhian Thought* (Delhi: Ambe Books), 1995.

Karan Singh, *Prophet of Indian Nationalism* (Bombay: Bhartiya Vidya Bhawan), 1991.

Karunakaran, K.P. (ed.), *Modern Indian Political Thought* (New Delhi: Allied Publishers), 1962.

Kaushik, P.D., *The Congress Ideology and Programme, 1920-47* (New Delhi: Allied), 1964.

Ker, James Compbell, *Political Trouble in India* (Calcutta: Superintendent Government Printing), 1917.

Krishandas, *Seven Months with Mahatma Gandhi* (Madras: Ganesan), 1928.

Kothari, Rajni, *Politics in India* (Boston: Little Brown and Co.), 1970.

Laski, Harold, J., *A Grammar of Politics* (London: Oxford University Press), 1970.

Le Fanu, *A Manual of Salem District* (Madras: Government Press), 1883.

Lovett, Sir Verney, *A History of Indian Nationalist Movement* (London: John Murray), 1921.

Mahadev Desai's Diary, Vols. 1 to 8 (Varanasi: Sarva Seva Sangh).

Majumdar, R.C., *History of the Freedom Movement in India* (Calcutta: Firma K.L. Mukhopadhyaya), 1963.

Malleson, G.B. and Kye John, *History of Indian Mutiny* (London: Allen), 1879.

Masani, R.P., *Dadabhai Naoroji* (New Delhi: Publications Division, Ministry of Information and Broadcasting, Government of India), 1960.

Masselos, Jim, *Indian Nationalism: An History* (New Delhi: Sterling), 1985.

Mathur, B.D., *Gokhale, A Political Biography* (Bombay: Himalaya Publishers), 1955.

McCarthy, Justin, *History of Our Times*, Vol. II (London: Chatts), 1897.

Mead, Henery, *The Sepoy Revolt: Its Causes & Consequences* (Delhi: Gyan), 1986.

Mende, Tibor, *Conversations with Mr. Nehru* (Bombay:Asia Publishing House), 1985

Menon, V.P., *An Outline of Indian Constitutional History* (Bombay: Bhartiya Vidya Bhavan), 1965.

Menon, V.P., *The Transfer of Power in India* (Bombay: Orient Longman), 1957.

Michael, Brecher, *Nehru—A Political Biography* (London: London Press), 1959.

Mody, Homi, *Pherozeshah Mehta* (New Delhi: Publications Division, Ministry of Information and Broadcasting, Government of India), 1963.

Mukherjee, Radhakamal, *India: The Dawn of New Era* (New Delhi: Radha Publications), 1997.

Munshi, K.M. and R.R. Diwakar (eds.), *Hinduism Doctrine and Way of Life* (Bombay: Bhartiya Vidya Bhavan), 1970.

Nanda, B.R., *Gandhi, Pan-Islamism, Imperialism and Nationalism in India* (Bombay: Oxford University Press), 1989.

Nanda, B.R., *The Making of a Nation: India's Road to Independence* (New Delhi: Harper Collins India Pvt. Ltd.), 1998.

Narasimhan, V.K., *Kasturiranga Iyengar* (New Delhi: Publications Division, Ministry of Information and Broadcasting, Government of India), 1963.

Nehru, Jawaharlal, *An Autobiography* (London: The Bodley Head), 1958.

Nehru, Jawaharlal, *The Discovery of India* (New Delhi: Oxford University Press), 1999.

Nehru, Jawaharlal, *Selected Works* (New Delhi: Orient Longman), 1968.

Nehru, Jawaharlal, *India's Foreign Policy* (New Delhi: Orient Longman), 1969.

Payne, Robert, *The Life and Death of Mahatma Gandhi* (New Delhi: Sage Publishers), 1977.

Perumal, Nilkan, *Rajaji* (Calcutta: Maya Publications), 1948.

Pyarelal, *The Epic Fast* (Ahmedabad: Navajivan Publishing House), 1932.

Radhakrishanan, S. (ed.), *Mahatma Gandhi* (London: George Allen), 1949.

Rajagopalachari, C., *Articles in Hindustan Review*, Patna, June and July 1907.

Rajagopalachari, C., *Ambedkar Refuted* (Bombay: Hind Kitabs), 1946.

Rajagopalachari, C., *Chats Behind Bars* (Madras: S. Ganesan), 1931.

Rajagopalachari, C., *Hinduism: Doctrine and Way of Life* (Bombay: Bhartiya Vidya Bhavan), 1946.

Rajagopalachari, C., *Indian Prohibition Manual* (Madras: AICC), 1931.

Rajagopalachari, C., *A Jail Diary* (Madras: Rochouse Publishers), 1941.

Rajagopalachari, C., "M. K. Gandhi: His Message to India", *Indian Review,* Madras, May 1916.

Rajagopalachari, C., *Plighted Words, Servants of Untouchables Society,* Delhi, 1933.

Rajagopalachari, C., *Satyam Eva Jayate* (Madras: Bharathan)

Rajagopalachari, C., Upanishads, *Hindustan Times*, Delhi, 1937.

Rajagopalachari, C. and Navaratna Rama Rao, *Our College: Reminiscences, undated article supplied by Central College,* Bangalore.

Rajaji (Madras: Bharathan), 1971.

Rajaji (Madras: Hindustan Publications), 1949.

Raju, P.T., *Idealistic Thought in India* (Cambridge: Harvard University Press), 1953.

Rao, B. Shiva, *India's Freedom Movement* (New Delhi: Longman), 1972.

Reddy, P.L. Sanjeev and Mishra, S.N. (eds.), *Sardar Vallabhbhai Patel. The Nation Builder* (New Delhi: Indian Institute of Public Administration), 2004.

Report of College Day Celebrations, Central College Old Boys' Association, Bangalore, 1937.

Richards, F.J., *Salem* (Madras: Government Press), 1918.

Salem District Urban Bank, Silver Jubilee Souvenir, 1934.

Sardar Patel's Correspondence (Ahmedabad: Navajivan Publishing House), 1973.

Shankhdher, M.M., *Reflections on Indian Politics* (New Delhi: Kumar Brothers), 1973.

Sharma, Jai Narain, *Power, Politics & Corruption: A Gandhian Solution* (New Delhi: Deep and Deep), 2004.

Shraddhanand, Swami, *Inside Congress* (Bombay: Phoenix), 1946.

Shukla, B.D., *Conversations of Gandhi* (Bombay: Vora Publishers), 1949.

Shukla, B.D., *History of the Indian Liberal Party* (Calcutta: Bharti Publishers), 1957.

Sitaramayya, P., *History of the Congress 1885 to 1936* (Bombay: Padama Publications), 1948.

Sitaramayya, Pattabhi, *History of Indian National Congress* (Delhi: S. Chand & Co.), 1969.

Stephens, Ian, *Monsoon Morning* (London: Ernest Benn), 1966.

Suda, J.P., *Main Currents of Social and Political Thought in India*, Vol. III (Meerut: K. Nath & Co.), 1973.

Tahamankar, *Lokmanaya Tilak* (London: John Murray), 1956.

Tara Chand, *History of Freedom Movement*, Vol. 3 (New Delhi: Publications Division, Information and Broadcasting, Government of India), 1972.

Tendulkar, D.G., *Mahatma*, 8 Vols. (New Delhi: Publications Division, Ministry of Information and Broadcasting, Government of India), 1958.

The Collected Works of Mahatma Gandhi, 100 Vols. (New Delhi: Government of India, Publications Division), 1958-1995.

The Calendar of 1898-99, University of Madras, 1898.

The 1921 Movement (New Delhi: Publications Division, Ministry of Information and Broadcasting, Government of India), 1971.

Varma, V.P., *Modern Indian Political Thought* (Agra: Luxmi Narain Agarwal), 1961.

Vajpeyi, J.N., *The Extremist Movement in India* (Allahabad: Chug Publications), 1974.

Wardlow, Grant, *Political Terrorism Theory, Tactics and Counter Measures* (Cambridge: Cambridge University Press), 1982.

Wilson, F.W., *The Indian Chaos* (London: London Press), 1932.

Zaidi, A.M. (ed.), *The Encycolpaedia of Indian National Congress* (New Delhi: S. Chand), 1979.

Index